6G AND ONWARD TO NEXT G

6G AND ONWARD TO NEXT G

THE ROAD TO THE MULTIVERSE

Martin Maier

Institut National de la Recherche Scientifique (INRS)
Montréal, Canada

IEEE
COMMUNICATIONS
SOCIETY

The ComSoc Guides to Communications Technologies
Nim K. Cheung, *Series Editor*
Richard Lau, *Associate Series Editor*

IEEE PRESS

WILEY

Published by John Wiley & Sons, Inc., Hoboken, New Jersey.
Published simultaneously in Canada.

For general information on our other products and services or for technical support, please contact our Customer Care Department within the United States at (800) 762-2974, outside the United States at (317) 572-3993 or fax (317) 572-4002.

Wiley also publishes its books in a variety of electronic formats. Some content that appears in print may not be available in electronic formats. For more information about Wiley products, visit our web site at www.wiley.com.

Library of Congress Cataloging-in-Publication Data Applied for:

Hardback ISBN: 9781119898542

Cover Design: Wiley
Cover Image: © TogsDesign/Shutterstock

Set in 10/12pt JansonTextLTStd by Straive, Chennai, India

"We shape our tools and then our tools shape us ...
We become what we behold."

Marshall McLuhan
(1911–1980)

Canada's eminent media theorist and philosopher:

Credited with predicting the rise of the Internet and phrasing the
term global village in *The Gutenberg Galaxy*

CONTENTS

ABOUT THE AUTHOR

Maier is a full professor with the Institut National de la Recherche Scientifique (INRS), Montréal, Canada. He was educated at the Technical University of Berlin, Germany, and received both MSc and PhD degrees with distinctions (summa cum laude) in 1998 and 2003, respectively. He was a recipient of the two-year Deutsche Telekom doctoral scholarship from 1999 through 2001. In 2003, he was a postdoc fellow at the Massachusetts Institute of Technology (MIT), Cambridge, MA. He was a visiting professor at Stanford University, Stanford, CA, from 2006 through 2007. He was a corecipient of the 2009 IEEE Communications Society Best Tutorial Paper Award. Further, he was a Marie Curie IIF Fellow of the European Commission from 2014 through 2015. In 2017, he received the Friedrich Wilhelm Bessel Research Award from the Alexander von Humboldt (AvH) Foundation in recognition of his accomplishments in research on FiWi-enhanced mobile networks. In 2017, he was named one of the three most promising scientists in the category "Contribution to a better society" of the Marie Skłodowska-Curie Actions (MSCA) 2017 Prize Award of the European Commission. In 2019/2020, he held a UC3M-Banco de Santander Excellence Chair at Universidad Carlos III de Madrid (UC3M), Madrid, Spain. He is coauthor of the book "Toward 6G: A New Era of Convergence" (Wiley-IEEE Press, January 2021) and author of the book "6G and Onward to Next G: The Road to the Multiverse" (Wiley-IEEE Press, January 2023).

PREFACE

Global crises such as the Covid-19 pandemic highlighted the fragility of our current approach to globalized production, especially where value chains serve basic human needs. On the flip side, however, virtual experiences such as Zoom's cloud-based video platform for online meetings and events have skyrocketed in popularity as the Covid-19 pandemic's online-everything transformation took place. With the mass digital adoption of remote work and online social activities accelerated by a global pandemic, we may finally find ourselves on the verge of something big and potentially paradigm-shifting: *The Metaverse* – the next step after the Internet, similar to how the mobile Internet expanded and enhanced the early Internet in the 1990s and 2000s.

The Metaverse will be about being inside the Internet rather than simply looking at it from a phone or computer screen. With the rise of the Metaverse, the Internet will no longer be at arm's length. Instead, it will surround us and will radically reshape society. Importantly, the Metaverse is not only based on the social value of today's *Generation Z* that online and offline selves are not different; this is because the younger generation considers the social meaning of the virtual world as important as that of the real world since they think that their identity in virtual space and reality is the same. But it also aims at realizing the *fusion of digital and real worlds across all dimensions* created and delivered by non-traditional converged service platforms of future 6G and Next G networks, where developers do not hesitate to use technologies from as many disciplines as possible. They do not discriminate whether services and applications will be used by human beings or by physical, digital, or virtual objects.

The Metaverse, underpinned by decentralized Web3 technology, is widely seen as the precursor of the *Multiverse*. While the Metaverse primarily focuses on virtual reality (VR) and augmented reality (AR), the Multiverse offers eight advanced types of extended reality (XR) realms, which together span the entire reality–virtuality continuum, including but not limited to VR and AR. This book is a sequel to our last book titled "Toward 6G: A New Era of Convergence" (Wiley-IEEE Press, January 2021), in which we briefly touched on the Multiverse and argued that 6G should not only explore more spectrum at high-frequency bands but also, more importantly, converge driving technological trends such as multisensory XR applications, connected robotics and autonomous systems, wireless brain–computer interaction (a subclass of human–machine interaction), as well as blockchain and distributed ledger technologies.

The purpose of this book is to complement our prequel book by describing the most recent progress and ongoing developments in the area of the Metaverse and Multiverse. Specifically, the book aims at weaving the following *emerging themes* carefully together in future 6G and Next G networks and the enhanced services they offer to disruptive applications in order to enable peak-experiences and human transformations: (i) touch-screen typing will likely become outdated, while wearable devices will become commonplace, enabling future communication technologies that are anticipated to fold into our surroundings, thereby helping us get our noses off the smartphone screens and back into our physical and biological environments, (ii) human transformation through unifying experiences across the physical, biological, and digital worlds, (iii) seamless convergence and harmonious operations of communication and computation to provide user-intended services and change or even transform the behavior of humans through social influence, (iv) creation of new virtual worlds to create a mixed-reality, super-physical world that enables new superhuman capabilities, and (v) rise of a new regime that connects all humans and machines into a global matrix, which some call the global mind or world brain, leveraging on the collective intelligence of all humans combined with the collective behavior of all machines, plus the intelligence of nature, plus whatever behavior emerges from this whole.

Despite the current lack of compelling use cases and potential pitfalls of the Metaverse, a rising number of organizations are searching for ways to use the Metaverse. This book points to three spheres of contexts, in which we outline different narratives for the year 2030 and beyond. Due to their striking similarities, we select *Society 5.0* as the frame story or, if you will, *meta narrative*, in which the Metaverse as well as Multiverse can be embedded naturally. Taking into account our meta narrative as well as the fact that future 6G and Next G networks are anticipated to become more human-centered than previous generations of mobile networks, cross-disciplinary research is necessary, involving not only communications and computer science but also cognitive science, social sciences, psychology, and behavioral economics. In addition, as we shall see, neuroscience and psychological approaches should be used to better understand humans and thus build a *deeper Metaverse*. This book aims at providing the reader with new complementary material, putting a particular focus on 6G and Next G networks in the context of the emerging Metaverse as the successor of today's mobile Internet and precursor of tomorrow's Multiverse. We hope that this book will be instrumental in helping the reader find and overcome some of the most common 6G and Next G blind spots.

Montréal, 24 August 2022   *Martin Maier*

ACKNOWLEDGMENTS

First and foremost, I am deeply grateful to Dr. Nim Cheung, who invited me to write this sequel to our last Wiley-IEEE Press book titled "Toward 6G: A New Era of Convergence," in which Dr. Amin Ebrahimzadeh and I have explored the latest developments and recent progress on the key technologies enabling next-generation 6G mobile networks, ranging from autonomous AI agents and mobile robots to multi-sensory haptic communications and delivery of advanced XR experiences in a 6G post-smartphone era, while putting a particular focus on their seamless convergence. Further, I am grateful to Dr. Abdeljalil Beniiche and Dr. Sajjad Rostami for their contributions to the experimental results reported in Chapters 6, 7, and 8 of this book, as well as for drawing some of the illustrative figures. Likewise, I am thankful to the invaluable comments and ideas put forth by friends, colleagues, and anonymous reviewers, who are simply too numerous to mention here by name. At Wiley-IEEE Press, I would like to thank Mary Hatcher, Victoria Bradshaw, and in particular Teresa Netzler for their guidance throughout the whole process of preparing the book. Moreover, I would like to acknowledge the Natural Sciences and Engineering Research Council of Canada (NSERC) for funding our research through their Discovery Grant programme. Finally, and most importantly, I would like to express my love and gratitude to my beautiful wife Alexie and our two kids Coby and Ashanti Diva for sharing their digitally native Generation Z enthusiasm and hands-on experiences with the emerging Metaverse.

M. M.

ACRONYMS

3GPP	Third Generation Partnership Project
4G	fourth generation
5G	fifth generation
6G	sixth generation
ABI	application binary interface
ACC	access control contract
ACP	artificial societies, computational experiments, parallel execution
ADC	analog-to-digital converter
AGI	artificial general intelligence
AI	artificial intelligence
AI4Net	AI for communication network
ANN	artificial neural network
API	application programming interface
AR	augmented reality
ARG	alternate reality game
ARIB	Association of Radio Industries and Businesses
ATIS	Alliance for Telecommunications Industry Solutions
AV	augmented virtuality
AWS	Amazon Web Services
B5G	beyond 5G
BIoT	blockchain-based Internet of things
BS	base station
BUMMER	behaviors of users modified, and made into an empire for rent
CAS	complex adaptive system
CI	collective intelligence
CIA	Central Intelligence Agency
CIC	Central Intelligence Corporation
CoC	computation oriented communications
CoZ	Crowd-of-Oz
CPS	cyber-physical system
CPSS	cyber-physical-social system
CPU	central processing unit
DAC	digital-to-analog converter
DAO	decentralized autonomous organization
DApp	decentralized application
DL	deep learning

DLT	distributed ledger technologies
DNA	deoxyribonucleic acid
DNN	deep neural network
DSOC	decentralized self-organizing cooperative
DSP	digital signal processing
DWE	digital world experience
ECDSA	elliptic curve digital signature algorithm
EOA	externally owned account
EPON	Ethernet passive optical network
ERC	Ethereum Request for Comments
ESF	edge sample forecasting
ESP	extrasensory perception
ESPN	extrasensory perception network
ETH	ether
ETSI	European Telecommunications Standards Institute
EVM	Ethereum virtual machine
F5G	fifth generation fixed network
FBT	fitness-beats-truth
FCC	Federal Communication Commission
FG-NET	Focus Group Technologies for Network
FiWi	fiber-wireless
FN	future networks
f-NFT	fractionalized non-fungible token
FTTE	fiber-to-the-everywhere-and-everything
GDP	gross domestic product
GFT	Google Flu Trends
GMPG	Global Multimedia Protocol Group
GPS	global positioning system
GPT	general-purpose technology
GPU	graphics processing unit
HART	human–agent–robot teamwork
HCI	human–computer interface
HetNet	heterogeneous network
HIN	hyper intelligent networks
HIT	human intelligence task
HITL	human-in-the-loop
HMD	head-mounted display
HMI	human–machine interaction
HMN	Harmonized Mobile Networks
HO	human operator
HSI	human–system interface
HTC	Holographic-type communication

HTML	Hypertext Markup Language
I2V	invisible-to-visible
IA	intelligence amplification
ICT	information and communications technologies
IEN	intelligence-endogenous network
IFrame	inline frame
IMT	International Mobile Telecommunication
IoE	Internet of everything
IoT	Internet of things
IPFS	inter-planetary file system
ISAC	integrated sensing and communications
ISACC	integrated sensing, communications, and computing
ISG	Industry Specification Group
IT	information technologies
ITU	International Telecommunication Union
ITU-R	ITU-radiocommunication sector
ITU-T	ITU-telecommunication sector
JC	judge contract
KPI	key performance indicator
LED	light-emitting diode
LOS	line-of-sight
LTE-A	long-term evolution advanced
M2M	machine-to-machine
MEC	multi-access edge computing
MIMO	multiple-input and multiple-output
ML	machine learning
MMORPG	massively multiplayer online role-playing game
MPP	mesh portal point
MR	mixed reality
MTurk	mechanical turk
MU	mobile user
MWI	many-worlds interpretation
NbS	nature-based solutions
NDE	near-death experience
Net4AI	communication network for AI
NFT	non-fungible token
NGMN	next generation mobile networks
NG-OAN	next-generation optical access network
NG-PON	next-generation passive optical network
NOMA	Non-orthogonal multiple access
NSF	National Science Foundation
OFDM	orthogonal frequency division multiplexing

OLT	optical line terminal
ONU	optical network unit
OTT	over-the-top
OWC	optical wireless communication
P2P	peer-to-peer
PC	personal computer
PDA	perceive-decide-act
pHRI	physical human–robot interaction
PMN	perceptive mobile network
PON	passive optical network
PoW	proof-of-work
QIT	quantum information technology
QKD	quantum key distribution
QoE	quality-of-experience
QoS	quality-of-service
QR	quick response
RAN	radio access network
REN	resource efficient networks
RFID	radio-frequency identification
RGB	red green blue
RIS	reconfigurable intelligent surface
RPC	remote procedure call
SAGSIN	space-air-ground-sea integrated network
SDG	sustainable development goal
SDO	standard development organization
sHRI	social human–robot interaction
SLAM	simultaneous localization and mapping
STER	selflessness, timelessness, effortlessness, and richness
TBSN	trust-based secure networks
TDM	time division multiplexing
TOR	teleoperator robot
TRC	TRON request for comments
TV	television
UAV	unmanned aerial vehicle
URL	Universal Resource Locator
URLLC	ultra-reliable and low-latency communication
V2X	vehicle-to-everything
VC	venture capital
VPN	virtual private network
VR	virtual reality
VUCA	volatile, uncertain, complex, and ambiguous
WDM	wavelength division multiplexing

WiFi	wireless fidelity
WLAN	wireless local area network
WoZ	Wizard-of-Oz
WRC	world radiocommunication conference
XAI	explainable artificial intelligence
XR	extended reality

CHAPTER 1

Introduction

"Computers are useless. They can only give you answers."

PABLO PICASSO
(1881–1973)

1.1. Toward 6G: A New Era of Convergence

This book is a sequel to our last Wiley-IEEE Press book titled "Toward 6G: A New Era of Convergence," which was authored together with Amin Ebrahimzadeh and was the first published book on future 6G mobile networks [1].

In our prequel book, we argued that 6G should not only explore more spectrum at high-frequency bands but, more importantly, *converge driving technological trends*. Our applied approach was in line with the bold, forward-looking research agenda put forth by Saad et al. [2], which intends to serve as a basis for stimulating more out-of-the-box research that will drive the 6G revolution. Specifically, Saad et al. [2] claim that there will be the following four driving applications behind 6G: (i) multisensory extended reality (XR) applications, (ii) connected robotics and autonomous systems, (iii) wireless brain–computer interaction (a subclass of human–machine interaction), and (iv) blockchain and distributed ledger technologies. Among other 6G driving trends, they emphasize the importance of edge intelligence and the emergence of smart environments and new human-centric service classes, as well as the end of the smartphone era, given that smart wearables are increasingly replacing the functionalities of smartphones. They argue that smartphones were central to 4G and 5G. However, in recent years there has been an increase in wearable devices whose functionalities are gradually replacing those of smartphones, ranging from integrated headsets to smart body implants that can take direct sensory inputs from human senses.

These emerging smart wearables may bring an end to smartphones and potentially drive a majority of 6G use cases.

One of the most intriguing 6G visions out there at the time of writing our prequel book was outlined by Harish Viswanathan and Preben E. Mogensen, two Nokia Bell Labs Fellows, in an open access article titled "Communications in the 6G Era" [3]. In this article, the authors focus not only on the technologies but they also expect the human transformation in the 6G era through *unifying experiences across the physical, biological, and digital worlds* in what they refer to as the network with the sixth sense. Combining the multi-modal sensing capabilities with the cognitive technologies enabled by the 6G platform will allow for analyzing behavioral patterns and people's preferences and even emotions, hence creating a sixth sense that anticipates user needs and allowing for interactions with the physical world in a much more intuitive way.

Furthermore, Viswanathan and Mogensen [3] claim that new themes are likely to emerge. Specifically, the future of connectivity is in the creation of *digital twin worlds* that are a true representation of the physical and biological worlds at every spatial and time instant, unifying our experience across these physical, biological, and digital worlds. Digital twins of various objects created in edge clouds will form the essential foundation of the future digital world. Digital twin worlds of both physical and biological entities will be an essential platform for the new digital services of the future. Digitalization will also pave the way for the creation of new virtual worlds with digital representations of imaginary objects that can be blended with the digital twin world to various degrees to create a mixed-reality, super-physical world, enabling new *superhuman* capabilities. Augmented reality (AR) user interfaces will enable efficient and intuitive human control of all these worlds, whether physical, virtual, or biological, thus creating a unified experience for humans and the human transformation resulting from it. Dynamic digital twins in the digital world with increasingly accurate, synchronous updates of the physical world will be an essential platform for augmenting human intelligence.

Importantly, Viswanathan and Mogensen [3] outlined a *vision of the future life and digital society* on the other side of the 2030s. While the smartphone and the tablet will still be around, we are likely to see new man–machine interfaces that will make it substantially more convenient for us to consume and control information. The authors expect that wearable devices, such as earbuds and devices embedded in our clothing, will become common. We will have multiple wearables that we carry with us and they will work seamlessly with each other, providing natural, intuitive interfaces. Touch-screen typing will likely become outdated. Gesturing and talking to whatever devices we use to get things done will become the norm. The devices we use will be fully context-aware, and the network will become increasingly sophisticated at predicting our needs. This context awareness combined with new man–machine interfaces will make our interaction with the physical and digital world much more intuitive and efficient. The computing needed for these devices will likely not all reside in the devices themselves because of form factor and battery power considerations. Rather, they may have to rely on locally available computing resources to complete tasks beyond the edge cloud. As consumers, we can expect that the self-driving concept cars of today will be available to the masses by the 2030s. They will be self-driving most of the time and thus will substantially increase the time available for us to consume data from the Internet in the form of more entertainment, rich communications, or education. Further, numerous domestic service robots will complement the vacuum cleaners and lawn mowers we know today. These may take the form of a swarm of smaller robots that work together to accomplish tasks.

In fact, according to [4], nothing has happened yet in terms of the Internet. The Internet linked humans together into one very large thing. From this embryonic net will be born a collaborative interface, a sensing, cognitive apparatus with power that exceeds any previous invention. The hard version of it is a future brought about by the triumph of a superintelligence. According to Kelly, however, a soft singularity is more likely, where artificial intelligence (AI) and robots converge – humans plus machines – and together we move to a complex interdependence. This phase has already begun. We are connecting all humans and all machines into a global matrix, which some call the *global mind* or *world brain*. It is a new regime wherein our creations will make us better humans. This new platform will include the collective intelligence (CI) of all humans combined with the collective behavior of all machines, plus the intelligence of nature, plus whatever behavior emerges from this whole. Kelly estimates that by the year 2025 every person will have access to this platform via some almost-free device.

Our prequel book described the latest developments and recent progress on the key technologies enabling 6G mobile networks, paying particular attention to their seamless convergence. Among other potential research directions, 6G will take cloud services to the next level by moving many of the computational and storage functions from the smartphone to the cloud. As a result, most of the computational power of the smartphone can focus on presentation rendering, making virtual reality (VR), AR, or XR more impressive and affordable. Furthermore, 6G will transform a transmission network into a computing network. One of the possible trademarks of 6G could be the seamless convergence and harmonious operations of transmission, computing, AI, machine learning, and big data analytics such that 6G is expected to detect the users' transmission intent autonomously and automatically provide personalized services based on a user's intent and desire.

In the final chapter of our prequel book, we took an outlook on how future profound 6G technologies will weave themselves into the fabric of everyday life until they are indistinguishable from it. As a result, the boundary between virtual (i.e. online) and physical (i.e. offline) worlds is to become increasingly imperceptible, while both digital and physical capabilities of humans are to be extended via edge computing variants with embedded AI capabilities. More specifically, we elaborated on the implications of the transition from the current gadgets-based Internet to a future Internet that is evolving from bearables (e.g. smartphone), moves toward wearables (e.g. Google and Levi's smart jacket or Amazon's announced voice-controlled Echo Loop ring, glasses, and earbuds), and then finally progresses to nearables (e.g. intelligent mobile robots). Nearables denote nearby surroundings or environments with embedded computing/storage technologies and service provisioning mechanisms that are intelligent enough to learn and react according to user context and history in order to provide user-intended services. While 5G was supposed to be about the Internet of Everything (IoE), to be transformative 6G might be just about the opposite of Everything, i.e. Nothing or, more technically, No Things. Toward this end, we introduced the *Internet of No Things* as an extension of immersive VR from virtual to real environments, where human-intended Internet services – either digital or physical – appear when needed and disappear when not needed. In doing so, the Internet of No Things helps tie both online and offline worlds closer together for the extension of human capabilities and experiences, ranging from conventional VR and AR to advanced XR and even more sophisticated cross-reality environments that involve various types of physical and digital realities.

Figure 1.1 depicts our proposed architecture of the Internet of No Things, which integrates the following three evolutionary stages of mobile computing: (i) ubiquitous, (ii) pervasive, and (iii) persuasive computing. Ubiquitous computing is embedded in the things surrounding us (i.e. nearables), while pervasive computing involves bearables and wearables. Persuasive computing aims at changing or even transforming the behavior of human users through social influence. As explained in technically greater detail in Chapter 5, the Internet of No Things will be instrumental in not only establishing XR as the next-generation mobile computing platform for the extension of human capabilities and experiences but also enabling future communication technologies that are anticipated to fold into our surroundings, thereby helping us get our noses off the smartphone screens and back into our physical and biological environments. The Internet of No Things represents an important stepping stone toward ushering in the 6G post-smartphone era and its underlying fusion of digital and real worlds created and delivered by non-traditional converged service platforms.

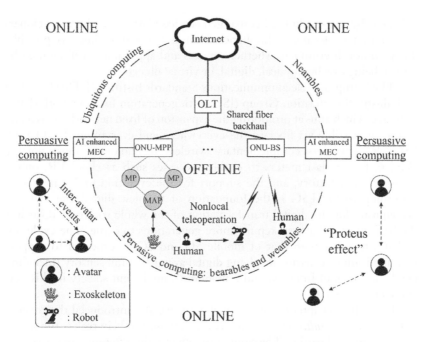

Figure 1.1 Internet of No Things: Integrating ubiquitous, pervasive, and persuasive computing for the extension of human capabilities and experiences.
Source: Maier et al. (2020). © 2020 IEEE.

1.2. Fusion of Digital and Real Worlds: Multiverse vs. Metaverse

In May 2019, the ITU-T Focus Group Network 2030 (FG-NET-2030), an initiative focusing on the fixed networks domain, published the first white paper on their Network 2030 vision [5]. Network 2030 is an abstraction of network technologies required to deliver advanced applications in 2030 and the decade after. It aims at coexisting with deployed infrastructures, incrementally inserting new capabilities in both public and private fixed (wireline) networks. According to [5], the next frontier in multimedia after VR and AR will include holographic media and multi-sense network services, e.g. haptic communication services. Soon our experiences with VR/AR will determine that they are not real enough, calling for new media, unencumbered by today's head-mounted displays (HMDs). The fusion of digital and real worlds across all dimensions is the driving theme for Network 2030, created and delivered by non-traditional converged service platforms, where developers do not hesitate to use technologies from as many disciplines as possible. They do not discriminate whether services and applications will be used by human beings, or by physical, digital, or virtual objects.

The European Telecommunications Standards Institute (ETSI) launched its Industry Specification Group (ISG) fifth generation fixed network (F5G) initiative, which aims at promoting the expansion of fixed networks to as many sectors as possible via fiber-to-the-everywhere-and-everything (FTTE) [6]. F5G also considers complementary wireless technologies, most notably WiFi 6, for the last meters to enable use cases such as cloud VR, online gaming, smart factory, and the support for the evolution of 5G networks. According to [6], F5G is the foundation of the new digital age and is a prerequisite for the digital transformation of the whole society. F5G is just the beginning and a first step for more generations to come. The evolution of F5G, together with that of mobile 5G and 6G, is expected to support new application scenarios involving digital avatar life, full sensory (including tactile and haptic) Internet, and a ubiquitous intelligent society in this new era of convergence.

In the final chapter of our prequel book, we also introduced the concept of the so-called *Multiverse* as an interesting attempt to help realize the fusion of digital and real worlds. The Multiverse offers eight different types of reality, including but not limited to VR and AR, as explained shortly. A term closely related to the Multiverse is the recently emerging *Metaverse*. According to [7], the Metaverse will be the precursor of the Multiverse. Specifically, the Metaverse might be viewed as the next step after the Internet, similar to how the mobile Internet expanded and enhanced the early Internet in the 1990s and 2000s. The various adventures that this place has to offer will surround us both socially and visually. The Metaverse is unique in that it spans a wide range of interconnected platforms as well as the digital and physical worlds underpinned by decentralized *Web3* technology.

As shown in Figure 1.2, while the Web1 (read-only web) and Web2 (read-and-write web) enabled the knowledge economy and today's platform economy, respectively, the Web3 will enable the *token economy* where anyone's contribution is compensated with a token. The token economy enables completely new use cases, business models, and types of assets and access rights in a digital way that were economically not feasible before, thus enabling completely new use cases and value creation models. Note that the term token economy is far from novel. In cognitive psychology, it has been widely studied as a medium of exchange, and arguably more importantly, as a positive reinforcement method for establishing desirable human behavior, which in itself may be viewed as one kind of value creation. Unlike coins, however, which have been typically used only as a payment medium, tokens may serve a

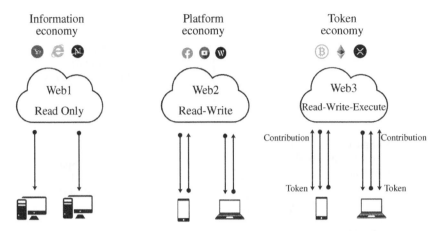

Figure 1.2   Evolution of Internet economy: From read-only Web1 information economy and read-write Web2 platform economy to read-write-execute Web3 token economy based on decentralized blockchain technologies.

wide range of different non-monetary purposes. Such purpose-driven tokens are instrumental in incentivizing an autonomous group of individuals to collaborate and contribute to a common goal. The exploration of tokens, in particular different types and roles, is still in the very early stages [8].

According to [7], the Metaverse will put the user first, allowing every member of our species to delve into new realms of possibilities. A modern, digital renaissance is taking place on the grandest state we have ever seen, involving billions of connected brains. In the coming decades, a new era of virtual life will bring in our next big milestone as a networked species.

In the following, we briefly elaborate on the salient features and main characteristics of the Multiverse and Metaverse.

1.2.1. The Multiverse: An Architecture of Advanced XR Experiences

In this section, we briefly highlight how the Multiverse can be used to tie both online and offline worlds closer together in the Internet of No Things. According to [9], the Multiverse offers a powerful experience design canvas to uncover hidden XR opportunities by fusing the real and the virtual, thereby creating *cross-reality environments* or so-called *third spaces*. Third spaces are created whenever one transverses the boundary between different XR realms within any given experience, as explained in more detail shortly. It is worthwhile to mention that, in "The Computer for the 21st Century,"

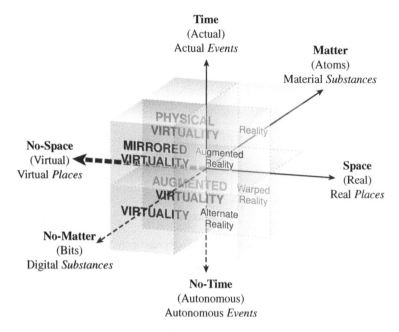

Figure 1.3   The Multiverse as an architecture of advanced XR experiences: Three dimensions, six variables, and eight realms.
Source: (Pine and Korn, 2011) © Berrett-Koehler Publishers.

Mark Weiser seems to had something similar in mind when describing what he initially called *embodied virtuality*, which is now more widely referred to as ubiquitous computing [10].

Apart from conventional VR and AR, future XR technologies may realize novel, unprecedented types of reality. Thus, X may be rather viewed as a placeholder for future yet unforeseen developments on the digital frontier. An interesting attempt to charter the unknown territory is the Multiverse, which may serve as an architecture of advanced XR experiences. As shown in Figure 1.3, the Multiverse consists of the following architectural components:

- **Dimensions:** There are the three well-known physical dimensions – Space, Time, and Matter – that constitute our physical reality.
- **Variables:** In addition, there are three non-physical dimensions – referred to as *No-Space, No-Time,* and *No-Matter* – that make up the virtual world. Unlike their physical counterparts, these three digital dimensions are not subject to the constraints imposed by physical space, time, and matter. Thus, in total there are six variables that can be exploited for the design of advanced XR experiences.

▪ **Realms:** Given that there are three (3) pairs of variables, each with two (2) opposite physical/digital dimensions, we have a total of $2^3 = 8$ possible realms. Each realm creates a different type of reality, ranging from conventional VR and AR to more sophisticated types of reality, e.g. mirrored virtuality, warped reality, and alternate reality. Mirrored virtuality absorbs the real world into the virtual and creates a virtual expression of reality that unfolds as it actually happens, providing a particular bird's eye view. Warped reality plays with time in any way possible by taking an experience firmly grounded in reality and shifting it from actual to autonomous time. Alternate reality, on the other hand, creates an alternative view of the real world by constructing a digital experience and superimposing it onto a real place. Unlike AR, however, alternate reality manipulates time and allows looking to the future freed from the bonds of actual time.

In Chapter 3, we will describe the Multiverse's eight different realms in technically greater detail and its great potential for the design of advanced XR experiences in a more comprehensive manner.

1.2.2. Metaverse: The Next Big Thing?

Recall from above that the Metaverse is anticipated to be the next Internet. The Metaverse is a new realm that will combine the actual and virtual worlds. It is all about virtual experiences and digital assets. Among others, the Metaverse ought to have the following main characteristics: (i) It must be a *shared experience*; just as it is in the real world, we get to witness events as they unfold; (ii) it must be possible to purchase and sell things to each other in a *virtual economy*; (iii) it has to be possible for people to participate in *activities that combine the real and virtual worlds*. Accordingly, the Metaverse has been described as a set of virtual experiences, locales, and products that increased in popularity as the Covid-19 pandemic's online-everything transformation took place. The Metaverse has the potential to alter practically every area of our life drastically. For instance, in the Metaverse, we can travel, study, work, consume entertainment, shop, and communicate with others. More importantly, the Metaverse will open up new avenues for earning a living and compensating for a broad and diversified spectrum of previously unrewarded creative activity (see also Web3 token economy above) [7].

Several companies have already embraced the Metaverse. Apple, Google, Samsung, The Walt Disney Company, Nintendo, Nvidia, Facebook, Amazon, Microsoft, Epic Games, and others are involved. For instance, recently, on 18 January 2022, Microsoft has announced the acquisition of game developer and interactive entertainment content publisher Activision Blizzard in an all-cash deal worth almost USD$69 billion. The deal is Microsoft's largest acquisition in its 46-year history. The acquisition is being widely seen as a big bet to keep Microsoft competitive in the burgeoning Metaverse space. Their key point is that the Metaverse has the potential for social interaction, experimentation, entertainment, and, most importantly, profit. A rising number of organizations are searching for ways to use it. While other businesses are still figuring out what the term means, the Metaverse is already gaining traction in the gaming industry, with Epic Games and Roblox leading the charge. The two video gaming titans present a vision of what the Metaverse may be in terms of content and audience. For instance, Epic Games' *Fortnite* gave a virtual concert that drew over 12 million people.[1] At the same time, Roblox and Gucci collaborated to build a virtual Gucci Garden environment where limited-edition virtual bags were sold. One of the digital bags was sold for the equivalent of USD$4115, USD$800 more expensive than the physical counterpart. Epic Games is also providing far more than simply a practical on-ramp to its Metaverse-building efforts. Thousands of games use its *Unreal Engine*, the second largest independent gaming engine, which makes it simple to interchange assets, integrate experiences, and share user profiles [7].

On the other hand, there have been some critical voices recently surfacing about the lack of compelling use cases and potential pitfalls of the Metaverse. Perhaps most famously, Elon Musk, CEO of SpaceX and Tesla, poked fun at the Metaverse. On 21 December 2021, in an interview with *The Babylon Bee*,[2] he noted that he grew up being told not to sit too close to a television screen, as it was bad for his eyesight, quoting him: "I don't know if I necessarily buy into this Metaverse stuff. Sure, you can put a TV on your nose," mocking the suggestion that people would willingly wear VR/AR headsets for big chunks of their day and the idea that this would actually transport a person into a different world, "although people talk to me a lot about it – Web3." He added: "I think we're far from disappearing into the Metaverse. This sounds just kind of buzzword-y," though he acknowledged that he might be seen as rejecting the Metaverse in the same way many dismissed the Internet in its early days of 1990s: "There's some danger that that's the case. But I currently am unable to see a compelling Metaverse situation. I don't get it. Maybe I will, but I don't get it yet."

---

[1] It is worthwhile to mention that, according to [7], Fortnite was allegedly only a side project. The game was published quietly in 2017, and by May 2020, 350 million individuals had signed up for an account – up 100 million from March 2019 – with over 57 million active players. Fortnite made more than USD$1.2 billion in its first 10 months, making it the first free-to-play game to do so. When it first launched in 2018, its mobile app made USD$2 million each day.

[2] For the full interview, please visit: https://babylonbee.com/podcast/basic/276.

Even more critical about the Metaverse is Ethan Zuckerman, former director of the Center for Civic Media at MIT. In an article in *The Atlantic*,[3] Zuckerman argues that Facebook's recently presented Metaverse imagines futures that have been imagined a thousand times before. He claims that Facebook's Metaverse looks pretty much like they imagined one would like in 1994, when he together with his collaborator Daniel Beck were hoping to recreate the vision that Neal Stephenson had outlined in his 1992 book *Snow Crash*. He admits that they were both self-conscious enough to understand that Snow Crash took place in a dystopia, and that Stephenson was positing a beautiful virtual world because the outside world had become so bad that no one wanted to live in it. Zuckerman concludes that today's Metaverse creators are missing the point. The Metaverse isn't about building perfect virtual escape hatches – it's about holding a mirror to our own broken, shared world.

In Chapter 2, we will delve deeper into the original vision of the Metaverse, outlined by Neal Stephenson in his seminal book *Snow Crash*, and contrast it to the more familiar concept of cyberspace. Unlike cyberspace that resides entirely in virtuality, the Metaverse aims at connecting virtuality with reality, making it possible for people and other sentient beings, intelligent mobile robots, as well as software AI agents to communicate and interact in shared environments. Further, Chapter 2 will introduce and explain in technically greater detail the main attributes and key enabling technologies of the Metaverse, including so-called *non-fungible tokens (NFTs)*.

1.3. The Big Picture: Narratives for 2030 and Beyond

1.3.1. From IoT-Based Industry 4.0 to Human-Centric Cyber–Physical–Social Systems

The current fourth industrial revolution has been enabled through the Internet of Things (IoT) in association with other emerging technologies, most notably cyber-physical systems (CPS). CPS help bridge the gap between manufacturing and information technologies (IT) and give birth to the *smart factory*. This technological evolution enables *Industry 4.0* as a prime agenda of the High-Tech Strategy 2020 Action Plan taken by the government of Germany, the Industrial Internet from General Electric in the USA, and the Internet+ from China. Smart factories under Industry 4.0 have several benefits such as optimal resource handling, but also imply minimum human intervention in manufacturing [11].

---

[3] Ethan Zuckerman, "Hey, Facebook, I Made a Metaverse 27 Years Ago," The Atlantic, 29 October 2021.

When human beings are functionally integrated into a CPS at the social, cognitive, and physical levels, it becomes a so-called *cyber-physical–social system (CPSS)*, whose members may engage in cyber-physical–social behaviors that eventually enable *metahuman* beings with types of superhuman capabilities. CPSS belong to the family of future techno-social systems that by design still require heavy involvement from humans at the network edge instead of automating them away. A promising example of such human-centric CPSS is the aforementioned Internet of No Things, which we briefly introduced above in Section 1.1 and which we will describe in technically greater detail below in Chapter 5. In addition, we will elaborate on how human-centric blockchain technologies, most notably the emerging *decentralized autonomous organization (DAO)*, which has become a hot topic spawned by the rapid development of blockchain technologies in recent years [12], may be exploited to enable the heavy involvement of humans interacting with autonomous AI agents and robots.

For further information and a comprehensive up-to-date survey of the state of the art, challenges, and opportunities of CPSS, we refer the interested reader to [13].

1.3.2. Human-Centric Industry 5.0

In this section, we touch on the anticipated transition from today's technology-driven Industry 4.0 to tomorrow's *human-centric Industry 5.0* and its two visions of human-robot co-working and a more holistic bioeconomy based on the two mutually beneficial principles of digitalization and, more interestingly, *biologization*.

Recently, in January 2021, the European Commission released the first edition of their policy brief on Industry 5.0 [14]. Industry 5.0 will be defined by a re-found and widened purposefulness, going beyond producing goods and services for profit. A purely profit-driven approach has become increasingly untenable. In a globalized world, a narrow focus on profit fails to account correctly for environmental and societal costs and benefits. Further, crises such as the Covid-19 pandemic highlighted the fragility of our current approach to globalized production, especially where value chains serve basic human needs, e.g. healthcare. This wider purpose constitutes three core elements: (i) human-centricity, (ii) sustainability, and (iii) resilience.

One of the most important paradigmatic transitions characterizing Industry 5.0 is the shift of focus from technology-driven progress to a thoroughly human-centric approach. An important prerequisite for Industry 5.0 is that technology serves people, rather than the other way around, by expanding the capabilities of workers (up-skilling and re-skilling) with innovative technological means such as VR/AR tools, mobile robots, and exoskeletons.

Currently, two visions emerge for Industry 5.0. The first one is human–robot co-working, where humans will focus on tasks requiring creativity and robots will do the rest. The second vision for Industry 5.0 is bioeconomy, i.e. a holistic approach toward the smart use of biological resources for industrial processes [15]. The bioeconomy has established itself worldwide as a mainstay for achieving a sustainable economy. Its success is based on our understanding of biological processes and principles that help revolutionize our economy dominated by fossil resources and create a suitable framework so that economy, ecology, and society are perceived as necessary single entity and not as rivals. More specifically, biologization will be the guiding principle of the bioeconomy. Biologization takes advantage of nature's efficiency for economic purposes – whether they be plants, animals, residues or natural organisms. Almost every discipline shares promising interfaces with biology. In the long term, biologization will be just as significant as a cross-cutting approach as digitalization already is today. Biologization will pave the way for Industry 5.0 in the same way as digitalization triggered Industry 4.0. It is also obvious that the two trends – biologization and digitalization – will be mutually beneficial [16].

It is interesting to note that in [14], the authors also elaborate on the relation between the concepts of Industry 5.0 and *Society 5.0*. While both concepts are related in the sense that they refer to a fundamental shift of our society and economy toward a new paradigm, Society 5.0 is not restricted to the manufacturing sector but addresses larger social challenges based on the integration of physical and virtual spaces, which we have discussed above in Section 1.2. In the following, we further elaborate on the Society 5.0 vision.

1.3.3. Society 5.0

Society 5.0 is an initiative of the Fifth Science and Technology Basic Plan taken by the government of Japan to facilitate a human-centered approach that puts humans in the loop of today's CPS [17]. The human-centeredness of Society 5.0 was recently investigated in technically greater detail by Gladden [18], who describes the goal of Society 5.0 as the ability to create equal opportunities for all and to provide the environment that helps unleash the full potential of each individual. To do so, Society 5.0 will leverage on emerging information and communications technologies (ICT) to its fullest such that physical, administrative, and social barriers to each individual's self-realization are removed. Gladden [18] concludes that from an anthropological perspective, Society 5.0's inclusion of diverse non-human entities – most notably social robots and AI agents – as participants is nothing new, but instead something quite ancient, a return to the unpredictability, wildness, and continual encounters with the other that characterized

Societies 1.0 and 2.0, thanks to the prevalence of diverse non-human agency resulting from a heavy reliance on animals as key participants in society and the societies' religious and spiritual dimension. For illustration, Figure 1.4 depicts the transition from past to future societies and their co-evolution with industry [18–20].

The Industrial Revolution reduced the agricultural population from more than 90% to less than 5%. Similarly, the IT revolution reduced the manufacturing population from more than 70% to approximately 15%. The Intelligence Revolution of the 6G era will reduce the entire services population to less than 10%. Upon the question where will people go and what will they do then, Wang [21] gives the following answer: Gaming! Not leisure, but scientific gaming in cyberspace. Artificial societies, Computational experiments, and Parallel execution – the so-called ACP approach – may form the scientific foundation while CPSS platforms may be the enabling infrastructure for the emergence of intelligent industries. Everything will have its parallel avatar or digital twin in the cyberspace such that we can conduct numerous scientific games before any major decision or operation. This new, yet unknown CPSS-enabled connected lifestyle and working environments will eventually lead to high satisfaction as well as enhanced capacity and efficiency. Further, Wang [21] foresees that the Multiverse or parallel universes based on Hugh Everett's many-worlds interpretation (MWI) of quantum physics will become a reality in the age of complex spaces with intelligent industries. However, he warns that the capability of CPSS to collect tremendous energy from the masses through crowdsourcing in the cyberspace and then release it into the physical space can bring us both favorable and unfavorable consequences. Therefore, one of the critical research challenges is the human-centric construction of complex spaces based on CPSS.

Similar to Industry 4.0/5.0, Society 5.0 merges the physical space and cyberspace by applying not only social robots and embodied AI but also emerging technologies such as ambient intelligence, VR/AR, and advanced human–computer interfaces (HCI), in addition to our aforementioned CPSS example of the Internet of No Things. As shown in Figure 1.4, Society 5.0 will also exploit *bionics* and *robonomics*. Robonomics studies the sociotechnical impact of social human–robot interaction (sHRI) as well as blockchain technologies such as the DAO as well as cryptocurrencies – not only coins but also tokens – for the social integration of robots into human society [22].

It is important to note that Society 5.0 counterbalances the commercial emphasis of Industry 4.0. If the Industry 4.0 paradigm is understood as focusing on the creation of the smart factory, the Society 5.0 is geared toward creating the world's first *super smart society*. More interestingly, according to [23], Society 5.0 also envisions a paradigm shift from conventional monetary to future nonmonetary economies based on technologies that can

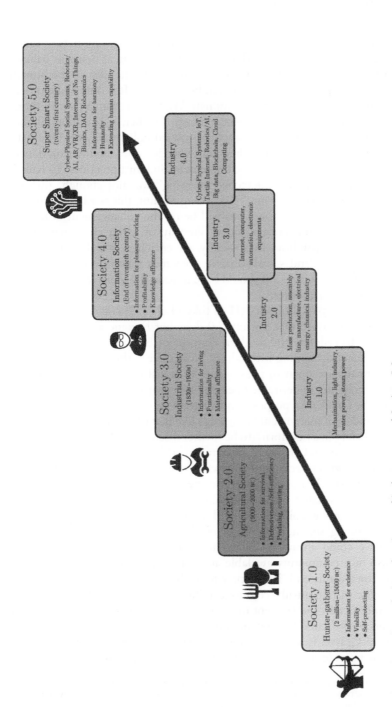

Figure 1.4 Co-evolution of society and industry toward Society 5.0.
Adapted from [18–20]. Beniiche et al. (2022). © 2022 IEEE.

measure activities toward *human co-becoming* that have no monetary value. As opposed to the Western traditional idea of the human as being or having, the idea that humans will be transformed equates to the idea of the human as becoming. We cannot become human by ourselves. It is only when others come to engage us that we become human. We become human with others. In other words, we are human co-becomings. Further, [23] elaborates that there are two paths in Buddhist practices toward enlightenment. One path changes one's mind, while the other path changes one's bodily experience. The two paths are sine qua non to complement Buddhist practices. To Kūkai, a Japanese Buddhist monk, "detached knowing" was not enough. Instead, he advocated "engaged knowing." According to him, this is what Buddhism is all about. In Society 5.0, the focus will be on the enhancement of human capabilities and the transformation of our way of living along with body and mind. Once capabilities in a society are enriched, social mobility will increase accordingly and social disparity becomes relatively weak. Toward this end, the future Society 5.0 should have indexes for social mobility as well as enrichment of capabilities.

Hitachi-UTokyo Laboratory (H-UTokyo Lab) [23] concludes by stating that in order to ensure that Society 5.0 does not become a dystopian society, we have to redefine the modern concept of humanity and find a path toward human co-becoming with others. Nonetheless, this path is not so easy, because humans are open to possibilities to transform themselves into any directions, including undesirable ones. In other words, we do not have a fixed *telos* (from the Greek *télos*, purpose, end, or goal) for co-becoming.[4] It would be wonderful indeed if our ancient knowledge like that of Kūkai turns up again in the future society in a new form (to be further discussed below in Section 1.4).

As we shall see in Chapter 8, in virtual worlds, people don't have trouble forming ad hoc groups constantly, because people are often thrown together by various quests. It is completely normal to walk up to strangers with no introduction whatever and ask them to join you in pursuit of a task. Virtual worlds unite collectivism and individualism in a complementary manner. In doing so, they create an *ideal community–individual relationship* in which we could always be independent if we wanted to be, but there would also be a community available at all times if we wanted to be part of a group. Toward this end, in Chapter 8, we borrow ideas from the *biological superorganism* with brain-like cognitive abilities observed in colonies of social insects. Specifically, the concept of stigmergy (from the Greek words *stigma* "sign" and *ergon* "work"), originally introduced in 1959 by French zoologist Pierre-Paul Grassé, is a class of self-organization mechanisms that made it possible to provide an elegant explanation to his paradoxical observations

---

[4]Télos is a term used by philosopher Aristotle to refer to the full potential or inherent purpose or objective of a person or thing, similar to the notion of an end goal or *raison d'être*. According to Wikipedia, it can be understood as the "supreme end of man's endeavour."

that in a social insect colony individuals work as if they were alone while their collective activities appear to be coordinated. In stigmergy, traces are left by individuals in their environment that may feed back on them and thus incite their subsequent actions. The colony records its activity in the environment using various forms of storage and uses this record to organize and constrain collective behavior through a feedback loop, thereby giving rise to the concept of *indirect communication*. As a result, stigmergy maintains social cohesion by the coupling of environmental and social organization. Note that with respect to the evolution of social life, the route from solitary to social life might not be as complex as one may think.

1.4. Purpose and Outline of Book

In our introductory discussion of 6G in Section 1.1, we have seen that next-generation mobile networks should not only explore more spectrum at high-frequency bands but, more importantly, converge 6G driving trends, most notably emerging smart wearables, smart environments, and new human-centric service classes. We have also seen that intriguing 6G visions foresee the emergence of the following new themes:

- **Theme 1:** Touch-screen typing will likely become outdated, while *wearable devices* will become common place, enabling future communication technologies that are anticipated to *fold into our surroundings*, thereby helping get our noses off the smartphone screens and back into our physical and biological environments.
- **Theme 2:** *Human transformation through unifying experiences* across the physical, biological, and digital worlds in what is referred to as the network with the sixth sense.
- **Theme 3:** Seamless convergence and harmonious operations of communication and computation to detect users' transmission intent autonomously and provide *user-intended services* by means of integrated ubiquitous, pervasive, and persuasive computing, aiming at changing or even transforming the behavior of humans through *social influence*.
- **Theme 4:** Creation of new *virtual worlds* that can be blended with the digital twin world to various degrees to create a mixed-reality, super-physical world that enables new *superhuman capabilities*.
- **Theme 5:** Rise of a new regime that connects all humans and machines into a global matrix, which some call the global mind or world brain, leveraging on the *CI* of all humans combined with the collective behavior of all machines, plus the *intelligence of nature*, plus whatever behavior emerges from this whole.

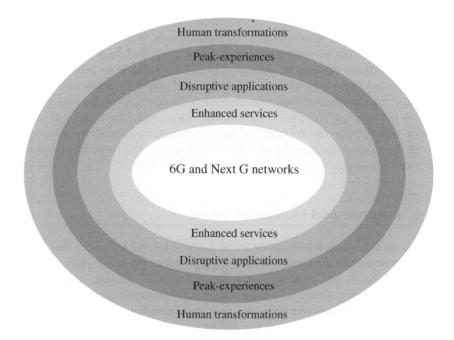

Figure 1.5   6G and Next G networks offering enhanced services to disruptive applications in order to enable peak-experiences and human transformations.

This book aims at weaving the aforementioned themes carefully together in future 6G and Next G networks and the enhanced services they offer to disruptive applications in order to enable peak-experiences and human transformations, as illustrated in Figure 1.5. Throughout the book, we pay particular attention to the fusion of digital and real worlds across all dimensions in the recently emerging Metaverse and the closely related Multiverse and its different types of reality, created and delivered by non-traditional converged service platforms, where developers do not hesitate to use technologies from as many disciplines as possible, including but not limited to technological disciplines. Recall from Section 1.2 that the Metaverse with its shared experience, virtual economy, and activities that combine the real and virtual worlds will be the precursor of the Multiverse. The Metaverse might be viewed as the next step after the Internet, which will surround us both socially and visually, underpinned by decentralized Web3 technology. The Web3 will enable the token economy where anyone's contribution is compensated with a token that may serve a wide range of different non-monetary purposes. We will elaborate on the token economy, a term widely studied in cognitive psychology for establishing desirable human behavior, in the context of 6G and Next G networks. Of particular interest will be purpose-driven tokens, which are instrumental in incentivizing an autonomous group of individuals to collaborate and contribute to a common goal and whose different types and roles are still in the very early stages of research.

We have seen that, despite the current lack of compelling use cases and potential pitfalls of the Metaverse, a rising number of organizations are searching for ways to use it. While other businesses are still figuring out what the term means, the Metaverse is already gaining traction in the gaming industry, with Epic Games' Fortnite and Roblox leading the charge. Due to this lack of compelling use cases, it is of critical importance to put the Metaverse in a bigger context in order to illustrate and thus help better understand its potential benefits. In Section 1.3, we have pointed to three spheres of contexts, in which we outlined different narratives for the year 2030 and beyond. In this book, we select Society 5.0 as the frame story or, if you will, *meta narrative*, in which the Metaverse as well as Multiverse can be embedded naturally. Using Society 5.0 as the meta narrative gives sense to the Multiverse and its precursor Metaverse, given that the Multiverse, Metaverse, and Society 5.0 bear striking similarities, as explained next.

1.4.1. Embedding the Multiverse and Metaverse in Meta Narrative Society 5.0

Recall from Section 1.3.3 that Society 5.0 has the following characteristics:

- **Goals:** Create world's first super smart society; create equal opportunities for all and provide the environment to unleash full potential of each individual; envision paradigm shift from conventional monetary to future non-monetary economies based on technologies that can measure activities toward human co-becoming; facilitate human transformation.
- **Approach:** Human-centered approach that puts humans in the loop of today's CPS; ACP approach forms scientific foundation for emergence of intelligent industries; scientific gaming with avatars or digital twins in cyberspace; Multiverse or parallel universes based on Hugh Everett's MWI of quantum physics will become reality in the age of complex spaces.
- **Enabling Technologies:** Leverage on emerging ICT to its fullest; CPSS platforms may be the enabling infrastructure for emergence of intelligent industries; apply emerging technologies such as social robots, embodied AI, ambient intelligence, VR/AR, advanced HCI, bionics, robonomics, DAO, and tokens.

- **Anthropological Perspective:** Inclusion of diverse non-human entities – most notably social robots, AI agents, avatars, and digital twins – as participants is nothing new, but instead something quite ancient
- **Religious and Spiritual Dimensions:** Apply Buddhist practices toward enlightenment such as "engaged knowing" advocated by Japanese Buddhist Kūkai; focus on enhancement of human capabilities.
- **Ancient Knowledge:** Télos is a term used by philosopher Aristotle to refer to the full potential or inherent purpose or objective of a person or thing, it can be understood as the supreme end of man's endeavor; humans do not have a fixed télos for co-becoming; it would be wonderful indeed if ancient knowledge like that of Buddhist Kūkai turns up again in future society in new form.
- **Metrics:** More complex metrics required to take social issues into account; develop indexes for enrichment of human capabilities as well as social mobility; once capabilities in society are enriched, social mobility increases and social disparity becomes relatively weak.
- **Risks:** Collecting tremendous energy from the masses through crowdsourcing in cyberspace and then release it into physical space can bring both favorable and unfavorable consequences; therefore, enable human-centric construction of complex spaces based on CPSS; redefine modern concept of humanity and find path toward human co-becoming with others to ensure that Society 5.0 does not become a dystopian society.

For illustration, Table 1.1 summarizes the above characteristics of Society 5.0, which we use as our meta narrative in which both Metaverse and Multiverse can be embedded naturally given their striking similarities and subtle differences. Recall from above that we'll describe the Metaverse and Multiverse in more depth in Chapters 2 and 3, respectively. Hence, some of the technical details shown in Table 1.1 will become clearer in these subsequent chapters. Notwithstanding, the table provides a useful overview that helps make the following insightful observations.

As shown in Table 1.1, super smartness and intelligence lie at the heart of both Society 5.0 and the Metaverse. Their common goal nicely aligns with the roadmap to 6G, outlined by Letaief et al. [24], which envisions that 6G will be transformative and will revolutionize the wireless evolution from "connected things" to "connected intelligence." As we will see in Chapter 4, 6G is anticipated to offer an ICT infrastructure that enables end users to perceive themselves as surrounded by a huge artificial brain providing humans with immense cognitive capabilities [25]. The resultant human transformations are

Table 1.1 Meta narrative Society 5.0 vs. embedded Metaverse and Multiverse: striking similarities and subtle differences.

	Meta narrative Society 5.0	Metaverse	Multiverse
Goals	Super smart society for twenty-first century	Central intelligence corporation in twenty-first century	The coming age of digital experiences
Approach	Humans put in loop of today's CPS	Humans paid for gathering intel	Human transformation via peak-experiences
Enabling Technologies	VR/AR, avatars, digital twins, social robots, embodied AI, CPSS, bionics, DAO, tokens	VR, avatars, daemons, hypercard, biological virus, metavirus, non-fungible tokens (NFTs)	VR/AR, XR, cross-reality environments, coins of time, eudaimonic technology
Anthropological Perspective	Nothing new, but instead something quite ancient: Inclusion of diverse non-human entities	At once brand new and very ancient: no difference between modern culture and Sumerian	First culture in history to use high technology to manufacture human experience, advent of postmaterialist culture
Religious and Spiritual Dimensions	Buddhist practices toward enlightenment, Japanese Buddhist Kūkai's "engaged knowing" for human co-becoming	Christian pentecostalism ("speaking in tongues"), Greek theomania, Shamanic "sacred" language of nature, "kinaturu" (African tongue of the ancestors of all magicians)	The Eternal: Realm of the truly Infinite, Eternity
Ancient Knowledge	Aristotle's concept of télos (367–347 BC)	Sumerian creation myth (5000 BC)	Limitless mind, emer-gence of consciousness
Metrics	More complex metrics required for advancement of human capabilities, social mobility, and equity	N/A	N/A
Risks	Dystopian society, crowdsourcing with un-favorable consequences	Dystopian future, neurolinguistic hacking, brain damage	Transversal experiences never before engendered or encountered

best achieved via peak-experiences, as illustrated above in Figure 1.5, which, unlike machines, humans thrive on and which are central not only to the coming age of digital experiences in the Multiverse, but also to the human-centric approach of Society 5.0 and the Metaverse. In Chapter 9, we will further elaborate on how the infinite potential of humans may be unleashed for our next evolutionary leap toward becoming metahuman. Given their underlying theme of fusion of digital and real worlds, it comes as no surprise that the Multiverse and Metaverse, embedded in the meta narrative Society 5.0, have many enabling technologies in common, ranging from VR/AR, avatars, social robots to DAO, (non-fungible) tokens, and *eudaimonic* technology in support of delivering the aforementioned peak-experiences.[5]

It is also interesting to note the overlap of religious and spiritual dimensions, rooted in ancient knowledge, whereby Society 5.0 and the Metaverse are nothing new, but instead something quite ancient. Or, put differently, they are *at once brand new and very ancient*, possibly ushering in the advent of a new culture in human history. Another important observation from Table 1.1 is the fact that neither the Metaverse nor the Multiverse define any specific metrics. This is where the choice of using Society 5.0 as meta narrative will be instrumental in defining more complex metrics required for measuring the advancement of human capabilities as well as social mobility and equity.

Finally, despite all the exciting opportunities on the road ahead to the Multiverse, it is of critical importance to address the risks of creating a dystopian society and a future with unintended negative consequences. Chapter 10 is therefore dedicated to the discussion of the opportunities and risks arising from the Metaverse.

1.4.2. The Art of 6G and Next G: How To Wire Society 5.0

Recently, in [26], we elaborated on how future 6G and Next G mobile networks should be applied to wire Society 5.0. Clearly, as described in our prequel book, open research challenges include millimeter-wave and terahertz (THz) communications, reconfigurable intelligent surfaces (RIS), the transition from network softwarization to network intelligentization, and the integration of underwater networks into four-tier space–air–ground–underwater network architectures. It is also worthwhile to briefly mention that many of the 5G key performance indicators (KPIs) will still be valid, though scaled by a factor of 10, 100, or even 1000, thus making 6G, in part, an incremental linear upgrade of 5G.

More importantly, we expanded on Alibaba Group's Yiqun Cai's recent OFC plenary talk on how technologies and applications drive the

---

[5]The term eudaimonia is etymologically based in the Greek words *eu* (good) and *daimon* (spirit). It is commonly translated as "happiness" or "welfare."

evolution of networking, where he concluded by presenting the equation *Networking = Art + Science + Engineering*. In the follow-up Q&A, he further elaborated on the first part of this equation, Art, by stating that modern networking requires not only scientific rigor and engineering know-how, but also creativity and originality. More specifically, he added that today's complex networks are more than computer science – they grow; they are life. Clearly, this implies that modern networks may be better viewed as techno-social systems that exhibit complex adaptive system (CAS) behavior and resemble biological superorganisms.

In the following, we briefly review the major conclusions drawn by Maier et al. [26], making the case that 6G and Next G should go beyond the incremental mindset of previous generations of mobile networks. Further, we highlight the unique potential benefits of the virtual world for society in that it provides a useful extension of the real-world economy by compensating for well-known market failures, e.g. rising income inequality.

1.4.2.1. Next G: Beyond Incremental 6G = 5G + 1G Mindset

There exists a wide consensus among the different stakeholders in the optical communications and networking community that there will be no 6G without fiber. To meet the very wide range of extreme requirements of beyond-5G (B5G) and 6G users, the latest progress on simplified digital coherent technologies for ultimate-capacity passive optical networks (PONs) carrying more than 100 Gbps per wavelength holds great promise for offering low-complexity optical access systems as well as key optical transport platforms heading toward 400 Gbps-based Ethernet transceivers for the 80 km range and over 800 Gbps short-reach applications [27].

The mobile radio spectrum, on the other hand, is not sufficient for the 100 Gbps data transmission rates envisaged for future 6G networks. According to [28], 6G requires a fundamental redesign of the physical layer extending beyond the theory of Shannon, giving rise to the so-called post-Shannon information theory with higher capacity potential. Furthermore, Fettweis and Boche [28] elaborate on possible true 6G innovations. Beside the symbiosis of radio communication and radio sensor technology as a key 6G technology, the authors argue that every odd-numbered generation of cellular networks first "practices" a new leap innovation with business users before the following even-numbered generation makes it a mass application for consumers. Specifically, while the promise of 5G is to start the Tactile Internet in order to control real and virtual objects in real time, 6G must provide an infrastructure to enable remotely controlled mobile robotic solutions for everyone – the Personal Tactile Internet.[6] Interestingly, the authors raise the important issue

---

[6]Note that this democratization of the Tactile Internet is identical to Bill Gates' original vision from the year 2007 of having a personal robot in every home [29].

of *trustworthiness* and postulate that the entire network architecture for 6G must be newly developed, which will require new procedures and paradigms. In fact, they claim that today's biggest challenge is the loss of trustworthiness and restoring trust must be understood as a basic societal challenge. For illustration, they draw a historical comparison between today and the renaissance, where Johannes Gutenberg's invention of the printing press in 1450 revolutionized society and heralded 300 years of renaissance. While some used the printing press to invent fake news and populism, Martin Luther used it to spread his ideas and overthrow the monopoly of knowledge of the monks.

Given the above challenges, one might argue that 6G should be more than only another cellular technology upgrade. Instead, 6G should go beyond continuing the linear incremental thinking 6G = 5G + 1G of past generations of mobile networks. The recently launched Next G Alliance is a bold new initiative to advance North American mobile technology leadership in 6G and beyond. Among its members are not only the major mobile network operators and manufacturers of North America but also, and arguably more interestingly, many of the key over-the-top (OTT) Internet players, including Apple, Google, Facebook, and Microsoft. One of the goals of the Next G Alliance is to create a roadmap that addresses the development and manufacturing across new markets and business sectors and promote widescale adoption of Next G technologies, both domestically and globally. To help get there, the current Next G roadmap sets audacious goals, most notably, (i) trust, security, and resilience, (ii) an enhanced digital world consisting of multi-sensory experiences, (iii) an AI native future network, while (iv) energy efficiency and environment must be at the forefront of decisions throughout the life cycle.

In our view, apart from meeting the traditional capacity and reliability requirements, one of the key 6G challenges of optical access systems and optical transport platforms is the support for precision of time in data delivery referred to as timeliness. Providing timeliness as a fundamental communication service will be central to the success of disruptive 6G applications. In particular, time-engineered communication services with coordinated guarantees over multiple traffic flows, which must be synchronized with respect to rendering of near real experiences, are anticipated to play a critical role in helping realize truly immersive *holographic-type communication (HTC)* for future hologram-based applications with skyrocketing volumetric bandwidth demands into the terabits-per-second range, an increase by several orders of magnitude over HD or even 3D VR video [30].

Unlike today's VR/AR HMDs, holograms will facilitate more natural experiences for human users and represent the ideal type of nearables to be inserted in the surrounding environment of our proposed Internet of No Things for the 6G post-smartphone era. Many hologram-based applications are highly interactive in nature and involve ultra-fast feedback loops, giving rise to the so-called "user interactivity challenge." Contrary to other types of multimedia services, the user interactivity challenge of immersive HTC will require ultra-low latency even if dealing with prerecorded content that does not involve real-time interaction with a remote party, as the user still interacts with the content simply by virtue of changing viewing angle and position. One specific challenge of minimizing network latency for HTC concerns the latency for the first packet in a flow. This generally incurs greater delay than later packets, as flow rules and entries may not yet have been installed. Clemm et al. [30] showed that *decentralization* of networks has a significant impact on reducing flow setup latency.

Clearly, advanced human-centered blockchain technologies such as validating on-chaining oracle and DAO hold great promise to resolve the aforementioned issues of trustworthiness and decentralization. Today's Internet is ushering in a new era. While the first generation of digital revolution brought us the Internet of information, the second generation – powered by decentralized blockchain technology – is bringing us the Internet of value, a true peer-to-peer platform that has the potential to go far beyond digital currencies and record virtually everything of value to humankind in a distributed fashion without powerful intermediaries. Arguably more importantly, though, the blockchain technology enables trusted collaboration that can start to change the way wealth is distributed as people can share more fully in the wealth they create. As a result, decentralized blockchain technology helps create platforms for distributed capitalism and a more inclusive economy [31].

While Gutenberg's invention gave birth to printing, the Internet's full potential still remains to be unleashed in the years to come. It is well known that Gutenberg's printing press played a pivotal role in Luther's reformation of society. According to [32], we don't yet know what the Internet truly is. Measured in Gutenberg time, we stand today at about the year 1481 with the progression of disruption in society. Note that Luther was born in the year 1483. Hence, the Internet's Martin Luther is yet to come.

1.4.2.2. Society 5.0: Exodus to the Virtual World?

In [33], Edward Castronova argues that people move their time and attention in response to their experiences in the separate realms of the real and virtual. He argues that the real and virtual worlds will become more similar. That is, the real world will become more like the virtual world, and vice versa.

Arguably more interesting, he states that it is indeed a singular power of the virtual world that it can create anything for free, except labor. It is considered absolutely intolerable that a player have nothing to do. Thus, the virtual world must ensure that there is always another quest to do and that every player, at all times, has some way of turning her own action into some reward. As a result, unlike the real world, the virtual world has the potential to provide full employment by design. What's more, the virtual world can start all players at zero wealth and anyone who needs money can get it, since work is always available. Hence, all players start on an equal basis, giving rise to *equality of opportunity*.

It has often been proposed that equality of opportunity ought to be the guiding principle of social policy. Policies that make the economic game obviously fairer are likely to become popular as virtual worlds broaden their influence. And what is striking is that even though online games exhibit economic inequality so vast that it dwarfs real-world inequality, nobody seems to care about inequality of outcomes. Indeed, Castronova observes that it is more fun if the outcomes actually do differ wildly, because players expect that if one acquires some new power she should also be offered greater rewards. Otherwise, the virtual world would be no fun at all. In other words, people don't complain about a lack of *vertical equity*, but they howl about failed *horizontal equity* – not only in the virtual world but also in the real world. Clearly, given its unique potential benefits, the virtual world provides a useful extension of the real-world economy by compensating for its well-known market failures, e.g. rising income inequality.

1.4.3. Finding Your 6G/Next G Blind Spot

We often use the term "blind spot" as a metaphor for the area of knowledge or understanding that we do not have or pay no attention to. For giving you the experience of the blind spot we humans have, Harris [34] used Figure 1.6 along with these instructions:

1. Hold this figure in front of you at arm's length.
2. Close your left eye and stare at the cross with your right.
3. Gradually move the page closer to your face while keeping your gaze fixed on the cross.
4. Notice when the dot on the right disappears.
5. Once you find your blind spot, continue to experiment with this figure by moving the page back and forth until any possibility of doubt about the existence of the blind spot has disappeared.

Figure 1.6 Visual experience of the existence of the blind spot: Follow the above instructions and see dot on the right disappearing and reappearing.

Punctuating ordinary experience in this way makes all the difference. Given this exemplary change in human perception of the world literally (not metaphorically) disappearing and reappearing in front of one's own eyes, this book aims at providing the reader with new complementary material that covers the latest Internet developments and that has not been covered in our prequel book. More specifically, this book puts a particular focus on 6G and Next G networks in the context of the emerging Metaverse as the successor of today's mobile Internet that has defined the last two decades. We hope that this book will be instrumental in helping readers inquire into new areas of knowledge or understanding that they didn't have or didn't pay attention to previously and thereby finding your 6G/Next G blind spot. Ideally, this book gives readers not only answers, but also confronts them with novel questions not asked let alone answered before.

1.4.4. Outline

The remainder of the book comprises the following 10 chapters:

In the first part of Chapter 2, we set the stage by providing a comprehensive description of the original Metaverse vision outlined by Neal Stephenson in his novel *Snow Crash*, which plays in Los Angeles in the twenty-first century, an unspecified number of years after a worldwide economic collapse and hyperinflation created by the government due to the loss of tax revenue as people increasingly began to use electronic currency. As we will see, snow crash is a hypercard which looks much like a business card and contains a *neurolinguistic virus* that uses the human brain as a host, jumping from one person to the next, kind of in the same way that a virus moves from one computer to another. Similar to machine language as the elemental language of computers, this neurological phenomenon comes from structures buried deep within the brain, common to all people, which can be found in many ancient cultures, sometimes referred to as the *language of nature*. The neurolinguistic virus is developed as a new powerful technology that is able to reprogram people's minds and alter their behaviors with the goal of advancing the human race by delivering it from the grip of the old civilization and its old rules people are stuck in. In the Metaverse, the key realization is that there's no difference between modern culture and ancient Sumerian. It is at once brand new and very ancient, getting us to observe the birth of a *new religion*. In the second part of Chapter 2, we highlight the different Metaverse components, applications, and open research challenges, followed by a couple of recent illustrative examples of early Metaverse deployments from both industry and academia. Among other open challenges, *cross-disciplinary research* will be necessary, involving cognitive science, social sciences, psychology, and economics, to better understand humans and thus build a deeper Metaverse.

Chapter 3 explores the infinite possibility of the Multiverse. Apart from describing each of its eight distinct realms of experience in technically greater detail, we elaborate on the importance of *experience innovation at the digital frontier* in business today, where people desire experiences – memorable events that engage people in inherently personal ways, emotionally, physically, intellectually, and even spiritually, more than the other economic offerings. More than in any of the other sectors of the economy (commodities, goods, and services), the currency that supports the emerging experience and transformation economies is the *coin of time*, where value will be determined more by how time is spent and less by how money is spent. We briefly introduce the *experience design canvas* as a valuable tool to take full advantage of the Multiverse and also discuss how its eight realms may be expanded outward in order to create *transversal experiences* never before envisioned, engendered, or encountered, which enable humans to do things they otherwise are not able to do. As a result, experiences are created that eventually help enable the transformation of humans. Finally, we touch on how the Multiverse goes beyond the Metaverse origins in that it creates third spaces that involve realms other than reality and virtuality, followed by some concluding remarks on the *limitless mind* and realms beyond time, space, and matter.

Next, in Chapter 4, we put the Metaverse and Multiverse in perspective of 6G and Next G networks. More specifically, we review the recent progress toward realizing the 6G vision as well as the current state of the art of 6G research activities. Among others, we discuss Ericsson's recent 6G research outlook and their anticipated *6G paradigm shifts*, e.g. from physical and digital worlds to a cyber-physical continuum, from data links to services beyond communication, and new capabilities, some of which may be more qualitative in nature. Moreover, we explain how future dual-functional wireless networks supported by integrated sensing and communications technologies act as the bond to bridge the physical and cyber worlds, giving rise to so-called *perceptive mobile networks (PMNs)*. We also touch on *quantum-enabled* 6G wireless networks, *blockchainized* mobile networks, and *AI-native* 6G networks. In our discussion, we pay particular attention to edge AI and *mimicking nature* to further imbue native intelligence by means of brain-inspired *stigmergy*, thereby enabling intelligent and seamless interactions among the human world, physical world, and digital world.

After presenting the 6G standardization roadmap, we further elaborate on the *difference between 6G and Next G* research. As we shall see, Next G research includes, but is not limited to, the specific KPI requirements and topics of interest addressed by 6G standard development organizations. Importantly, Next G research includes the Metaverse as one of their long-term objectives. Finally, we take a closer look at the recently published *Next G Alliance Roadmap* to 6G. Specifically, we outline Next G Alliance's audacious goals and priorities to address both the societal and economic needs, most notably the *acceleration of digital transformation* across society, *transformation of human interactions* across physical/digital/biological worlds, as well as human and machine 6G *Digital World Experiences (DWEs)* unthinkable with previous generations. Arguably more importantly, we also elaborate on the *symbiotic relationship* between technology and a population's societal and economic needs, whereby technology shapes human behavior and human needs shape technological evolution. Moreover, we explain the four foundational areas for 6G applications and use cases. According to Next G Alliance's Roadmap, humans are expected to be the ultimate beneficiaries of 6G, opening up new doors for development of *human-centric technologies* that positively influence human behaviors and technology-mediated human-to-human interactions, in turn advancing the societies they create. At the same time, *sustainability* must be at the forefront of decisions throughout the life cycle, given that climate challenges are expected to seriously disrupt business as usual and change the way citizens worldwide live their lives.

In Chapter 5, we elaborate on how the Internet of No Things with its underlying human-intended services may serve as a useful stepping stone toward realizing the far-reaching vision of future 6G networks, ushering in the 6G post-smartphone era, given that smart wearables are increasingly replacing the functionalities of smartphones. We then elaborate on the recently emerging *invisible-to-visible* technology concept, which we use together with other key enabling network technologies to tie both online and offline worlds closer together in an Internet of No Things and make it "see the invisible" through the awareness of nonlocal events in space and time. As an illustrative example of advanced XR experiences that transverse the boundary between the Multiverse realms, we study the delivery of *extrasensory human perceptions*, i.e. senses other than the five human senses. As we will see, with the advent of advanced XR technologies it might be easier to mimic the quantum realm instead of actually tapping into it. In addition, we present a *DAO use case study* where we explore how blockchain technologies, in particular the DAO, may be leveraged to decentralize the Tactile Internet as a promising example of future techno-social systems, which are anticipated to play an important role in 6G. In particular, we shed light on the importance of crowdsourcing of human expertise to solve problems that machines alone cannot solve well. Finally, we explore how the CI of unskilled crowd members of the DAO can be enhanced by means of nudging via a smart contract, giving rise to the concept of *hybrid-augmented intelligence* for addressing problems and requirements that may not be easily trained or classified by machine learning, especially due to fact that many problems that humans face tend to be of high uncertainty, complexity, and open-ended, and the Internet provides an immense innovation space for hybrid-augmented intelligence.

Chapter 6 explains Web3 and the important problem of *token engineering*, which is defined as the theory, practice, and tools to analyze, design, and verify tokenized ecosystems, in technically greater detail. Token engineering, also referred to as *token mechanism design*, is an emerging field. One may think of mechanism design as the engineering part of economic theory. It deals with the question of how to design a game, but starts at the end of the game (i.e. its desirable outcome) and then goes backward when designing the mechanism. Therefore, it is also referred to as "reverse game theory." As the field of tokens matures, it is likely that related disciplines like behavioral game theory will find its way into the modeling of tokens. Furthermore, this chapter describes the important process of *tokenization*, i.e. the process of creating tokenized digital twins via tokens. Tokens might be the killer application of Web3 networks. Specifically, we elaborate on the differences between fungible and NFTs, also known as NFTs, with regard to their interchangeability and divisibility. Of particular interest is the design of tokens, which may be programmed to have an expiration date with an inbuilt deflation (i.e. negative interest rate) to prevent hoarding and inflation, as exemplified by our discussion of "free money" (or Freigold) and Edison-Ford community money. Throughout Chapter 6, we pay close attention to the question of how so-called *purpose-driven tokens* may be used to create technology-enabled social organisms that enable the collective production of *tech-driven public goods* or *club goods*, which come with certain exclusion mechanisms in place, including associated *externalities* – both negative ones (costs) and positive ones (benefits). Overall, this chapter tries to help better understand how purpose-driven tokens are instrumental in restoring common goods and resolving many tragedy-of-the-commons problems society faces today by providing an *operating system for a new type of economy*.

In Chapter 7, we follow up on the previous chapter and introduce two examples of a new type of economy: *robonomics* and *tokenomics*. Robonomics is an emerging field, which studies the sociotechnical impact of blockchain technologies on behavioral economics and cryptocurrencies (both coins and tokens) for the social integration of robots into human society, including persuasive robotics as enforcer or supervisor of human behavior modification. Many studies have shown that the physical presence of robots benefits a variety of social interaction elements such as persuasion, likeability, and trustworthiness. Importantly, these robots are less like tools and more like partners, whose persuasive role in a social environment is mainly human-centric. We investigate the widely studied *trust game* of behavioral economics, an example of so-called public good games, in a blockchain context, which captures any generic economic exchange between two parties, while paying close attention to the importance of developing efficient cooperation and coordination technologies. We demonstrate experimentally that a social efficiency – a term closely related to *social capital* and *equity* – of up to 100% can be achieved by using the blockchain mechanism of deposit to enhance both *trust* and *trustworthiness*. We then present an *on-chaining oracle* blockchain architecture for a networked N-player trust game that involves a third type of human agents called observers, who track the players' investment and reciprocity. Of particular interest to us is the design of appropriate reward and penalty mechanisms. We experimentally demonstrate that the presence of third-party reward and penalty decisions helps raise the average normalized reciprocity above 80%, even without requiring any deposit. Further, we experimentally demonstrate that mixed logical-affective *persuasive strategies* for social robots improve the trustees' trustworthiness and reciprocity significantly. Finally, the chapter explains the anticipated paradigm shift from conventional monetary to future *nonmonetary economies* based on technologies that can measure activities toward human co-becoming that have no monetary value. Specifically, the chapter elaborates on the shift from conventional monetary economics to nonmonetary tokenomics enabled by tokenization in different value-based scenarios.

Chapter 8 delves into the *human-centeredness* of Society 5.0. While the primary focus of 5G has been on industry verticals, future 6G mobile networks are anticipated to become more human-centered. Toward this end, it is important to take a number of different factors into account in order to develop a more realistic understanding of *human nature* that challenges that of rational individuals driven by self-interest, as traditionally assumed in mainstream economics. Emerging CPSS aim at functionally integrating human beings into today's CPSs at the social, cognitive, and physical levels. CPSS are instrumental in realizing the human-centered Society 5.0 vision. Society 5.0 envisions human beings to increasingly interact with social robots and embodied AI in their daily lives. In this chapter, we expand on our work on robonomics and tokenomics in the previous chapter. After introducing our *CPSS based bottom-up multilayer token engineering framework for Society 5.0*, we experimentally demonstrate how the collective human intelligence of a blockchain-enabled DAO can be enhanced via purpose-driven tokens. More specifically, we aim at driving the *bionic convergence* of robonomics, DAO, and the Internet of No Things as our CPSS of choice to advance the CI of Society 5.0 in the next-generation Internet known as Web3. Importantly, we experimentally demonstrate the potential of the *biological stigmergy mechanism* for advancing CI in a CPSS-based DAO via tokenized digital twins, ushering in the future *stigmergic society* that will leverage on time-tested self-organization mechanisms borrowed from nature. Note that, in doing so, the future stigmergic society follows the guiding principle of *biologization*.

In Chapter 9, we double down on the potential of biologization as well as the mutually beneficial symbiosis between biologization and digitalization for the purpose of human development and our possible evolution into future *metahumans* with infinite capabilities. To better understand the natural potential of humans, we start by reviewing a recently proposed theory of biological human uniqueness and the qualities that make us humans and become such a distinctive species in comparison to our closest primate relatives with regard to (i) social cognition, (ii) coordinated decision-making, and (iii) uniquely human sociality. In particular, we show that evolution works only in response to a specific adaptive ecological problem that presents itself and how evolution makes those individuals best equipped to solve it have an adaptive advantage, thereby making us *smarter* and creating a *fundamentally new form of sociality* via a new form of cooperation and a concomitant new form of communication to support this cooperation, two attributes that are front and center in the super smart Society 5.0 vision. Leveraging on those insights into biological human uniqueness in a future human-centered Society 5.0, we then elaborate on their implications for the future Metaverse by introducing the concept of *symbiomimicry* as a promising means to help exit the Anthropocene and enter the Symbiocene in the coming 6G and Next G era. Further, given that the creation of novel shared physical and/or digital worlds in the Metaverse create the possibility of new kinds of concepts based on humans' unique capability of bifurcated experiences, we present a variety of different *concepts of the true nature of reality* that bring us one or more steps closer to the original Metaverse vision in Snow Crash, ranging from *MetaHuman Creator* for building a bespoke photorealistic digital human and *infinite reality* for realizing virtual immortality to *metareality* for harvesting transformative peak-experiences that hint at enormously expanded human potential.

Chapter 10 is dedicated to the discussion of the specific opportunities as well as risks of the Metaverse. Our comprehensive discussion of the Metaverse ranges from *cyberutopianism* to *digital cosmopolitanism*, while taking a close look at the recent end of globalization and the critical importance of technology, e.g. cryptocurrencies and decentralized network structures with *small-world* scale-free properties, for invoking the future of progress in a deglobalized world. We will elaborate on the crucial role decentralized small-world network structures play for designing methods to create robust networked systems, ranging from the social to the economic and the political, in order to avoid the collapse of whole systems.

Among other opportunities, we explain how the Metaverse may provide the digital tools for growing access to the *Global Mind*, whereby humans act as neurons in a *human hive mind* with blockchain technology acting as connective tissue to create virtual pheromone trails via programmable incentives and *extended stigmergy in dynamic media*. Furthermore, we explain how 6G, Next G, and the Metaverse may eventually pave the way to the *peak-experience machine* by democratizing access to the upper range of human experiences and making them available for the masses in order to foster mass flourishment and unleash the infinite potential of humans. On the flip side, we describe the risk that the Metaverse may turn humans into *cybernetic organisms (cyborgs)*, enhanced with internal neural implants to artificially create intellectual, emotional, and even spiritual experiences. The Metaverse may also become an advanced *behavior modification machine* that exploits a vulnerability in human psychology of uberworked and underpaid worker bees in an exacerbated platform capitalism. After weighing both opportunities and risks and outlining possible solutions, we wrap up Chapter 10 by envisioning a humanistic setting for the emerging Metaverse in support of *team human* in a future human-centric Society 5.0.

Finally, in Chapter 11, we draw our conclusions and provide an outlook on the future of 6G, Next G, and the road ahead to the Multiverse.

References

1. A. Ebrahimzadeh and M. Maier. *Toward 6G: A New Era of Convergence.* Wiley-IEEE Press, January 2021.
2. W. Saad, M. Bennis, and M. Chen. A vision of 6G wireless systems: applications, trends, technologies, and open research problems. *IEEE Network*, 34(3):134–142, May/June 2020.
3. H. Viswanathan and P. E. Mogensen. Communications in the 6G era. *IEEE Access*, 8: 57063–57074, March 2020.
4. K. Kelly. *The Inevitable: Understanding the 12 Technological Forces That Will Shape Our Future.* Viking, June 2016.
5. Focus Group Technologies for Network 2030 (FG-NET-2030). Network 2030 - A Blueprint of Technology, Applications and Market Drivers Towards the Year 2030 and Beyond. pages 1–19. ITU-T, May 2019.
6. ETSI. The Fifth Generation Fixed Network (F5G): Bringing Fibre to Everywhere and Everything. pages 1–24. White Paper No. #41, 1st edition, September 2020.
7. R. Higgins. *METAVERSE: A Definitive Beginners Guide to Metaverse Technology and How You Can Invest in Related Cryptocurrencies, NFTs, Top Metaverse Tokens, Games, And Digital Real Estate.* November 2021.
8. S. Voshmgir. *Token Economy: How the Web3 reinvents the Internet* (Second Edition). BlockchainHub, Berlin, Germany, June 2020.
9. B. J. Pine II and K. C. Korn. *Infinite Possibility: Creating Customer Value on the Digital Frontier.* Berrett-Koehler Publishers, August 2011.
10. M. Weiser. The computer for the 21st century. *Scientific American*, 265(3):94–104, September 1991.
11. D. Sinha and R. Roy. Reviewing cyber-physical system as a part of smart factory in industry 4.0. *IEEE Engineering Management Review*, 48(2):103–117, June 2020.

12. S. Wang, W. Ding, J. Li, Y. Yuan, L. Ouyang, and F.-Y. Wang. Decentralized autonomous organizations: concept, model, and applications. *IEEE Transactions on Computational Social Systems*, 6(5):870–878, October 2019.

13. Y. Zhou, F. R. Yu, J. Chen, and Y. Kuo. Cyber-physical-social systems: a state-of-the-art survey, challenges and opportunities. *IEEE Communication Surveys and Tutorials*, 22(1):389–425, Firstquarter 2020.

14. J. Cotta, M. Breque, L. De Nul, and A. Petridis. Industry 5.0: Towards a sustainable, human-centric and resilient European industry. pages 1–46, January 2021.

15. K. A. Demir, G. Döven, and B. Sezen. Industry 5.0 and Human-Robot Co-working. In *Proc., World Conference on Technology, Innovation and Entrepreneurship*, volume 158, pages 688–695. Procedia Computer Science, June 2019.

16. G. Schütte. What kind of innovation policy does the bioeconomy need? *New Biotechnology*, 40(Part A):82–86, January 2018.

17. Government of Japan. The 5th science and technology basic plan (tentative translation), December 2015. [Online; Accessed on 2022-06-09]. Available: https://www8.cao.go.jp/cstp/kihonkeikaku/5basicplan_en.pdf.

18. M. E. Gladden. Who will be the members of society 5.0? Towards an anthropology of technologically posthumanized future societies. *Social Sciences*, 8(5:148):1–39, May 2019.

19. N. Berberich, T. Nishida, and S. Suzuki. Harmonizing artificial intelligence for social good. *Philosophy & Technology*, 33: 613–638, December 2020.

20. M. Lammers. Society 5.0: Discover how proptech and the sustainable development goals come together, October 2020. [Online; Accessed on 2022-06-09]. Available: https://www.unissu.com/proptech-resources/Society-5-point-0-how-PropTech-and-the-Sustainable-Development-Goal-come-together.

21. F.-Y. Wang. The emergence of intelligent enterprises: from CPS to CPSS. *IEEE Intelligent Systems*, 25(4):85–88, July/August 2010.

22. I. S. Cardenas and J.-H. Kim. Robonomics: The Study of Robot-Human Peer-to-Peer Financial Transactions and Agreements. In *Proc., HRI' 20: Companion of the 2020 ACM/IEEE International Conference on Human-Robot Interaction*, pages 8–15, March 2020.

23. Hitachi-UTokyo Laboratory (H-UTokyo Lab). *Society 5.0: A People-Centric Super-Smart Society*. Springer Open, Singapore, May 2020.

24. K. B. Letaief, W. Chen, Y. Shi, J. Zhang, and Y. A. Zhang. The roadmap to 6G: AI empowered wireless networks. *IEEE Communications Magazine*, 57(8):84–90, August 2019.

25. E. Calvanese Strinati, S. Barbarossa, J. L. Gonzalez-Jimenez, D. Kténas, N. Cassiau, L. Maret, and C. Dehos. 6G: The next frontier: from holographic messaging to artificial intelligence using subterahertz and visible light communication. *IEEE Vehicular Technology Magazine*, 14(3):42–50, September 2019.

26. M. Maier, A. Ebrahimzadeh, A. Beniiche, and S. Rostami. The art of 6G (TAO 6G): how to wire society 5.0 [Invited]. *IEEE/OSA Journal of Optical Communications and Networking*, 14(2):A101–A112, February 2022.

27. N. Suzuki, H. Miura, K. Mochizuki, and K. Matsuda. Simplified digital coherent-based beyond-100G optical access systems for B5G/6G [Invited]. *IEEE/OSA Journal of Optical Communications and Networking*, 14(1):A1–A10, January 2022.

28. G. P. Fettweis and H. Boche. 6G: The personal tactile internet - and open questions for information theory. *IEEE BITS the Information Theory Magazine*, 1(1):71–82, September 2021.

29. B. Gates. A robot in every home. *Scientific American*, 296(1):58–65, January 2007.

30. A. Clemm, M. T. Vega, H. K. Ravuri, T. Wauters, and F. De Turk. Toward truly immersive holographic-type communication: challenges and solutions. *IEEE Communications Magazine*, 58(1):93–99, January 2020.

31. A. Beniiche, A. Ebrahimzadeh, and M. Maier. The way of the DAO: toward decentralizing the tactile internet. *IEEE Network*, 35(4):190–197, July/August 2021.

32. J. Jarvis. *Gutenberg the Geek*. Amazon Digital Services, February 2012.

33. E. Castronova. *Exodus to the Virtual World: How Online Fun is Changing Reality*. St. Martin's Griffin, November 2008.

34. S. Harris. *Waking Up: A Guide to Spirituality Without Religion*. Simon & Schuster, September 2014.

CHAPTER **2**

Metaverse: The New North Star

"Any sufficiently advanced technology is indistinguishable from magic."

Sir Arthur C. Clarke
Science-fiction writer, futurist, and inventor
(1917–2008)

The Metaverse is anticipated to be the successor to today's mobile Internet. Recall from Section 1.2.2 that the original vision of the Metaverse was outlined by Neal Stephenson in his novel *Snow Crash* published in 1992 [1]. According to Wikipedia, Stephenson's concept of the Metaverse has enjoyed continued popularity and strong influence in high-tech circles, especially Silicon Valley, ever since the publication of Snow Crash. As a result, Stephenson has become a sought-after futurist and has worked as a futurist for Blue Origin and, more recently, Magic Leap, an augmented reality (AR) startup.

Unlike cyberspace that resides entirely in virtuality, the Metaverse aims at connecting virtuality with reality. Note that similar to the Metaverse, the term cyberspace stems from a novel as well. In his 1984 science-fiction novel *Neuromancer*, William Gibson popularized the term cyberspace, which became the de-facto term for the World Wide Web during the 1990s [2]. It is interesting to note that the critically acclaimed movie *The Matrix* drew from Neuromancer the usage of the term "matrix" that depicts a virtual reality data space in a dystopian future, in which humanity is unknowingly trapped inside a *simulated reality* (to be revisited and further discussed in Chapter 9 in the context of the so-called simulation hypothesis).

In this chapter, we first describe the origins of the Metaverse in Snow Crash in a more comprehensive fashion. For illustration, we then highlight early deployments of the Metaverse in self-driving cars as well as other

6G and Onward to Next G: The Road to the Multiverse, First Edition. Martin Maier.
© 2023 The Institute of Electrical and Electronics Engineers, Inc.
Published 2023 by John Wiley & Sons, Inc.

use cases. Next, we explain in technically greater detail the key technologies enabling the Metaverse, including but not limited to the so-called non-fungible tokens (NFTs) briefly mentioned in Section 1.2.2.

2.1. Origins

To set the stage, Snow Crash plays in Los Angeles in the twenty-first century, an unspecified number of years after a worldwide economic collapse. The arrangements in Snow Crash resemble anarcho-capitalism. Hyperinflation was created by the government due to the loss of tax revenue as people increasingly began to use electronic currency. The novel's Central Intelligence Corporation (CIC) has emerged from the Central Intelligence Agency (CIA)'s merger with the Library of Congress. Contributors may gather intel and sell it to the CIC. It may be gossip, videotape, audiotape, a fragment of a computer disk, or a photocopy of a document. It can even be a joke based on the latest highly publicized disaster. Contributors are paid if their contributions are used by CIC's clients, mostly large corporations and Sovereigns. Users of the Metaverse gain access to it through personal terminals that project a high-quality virtual reality (VR) display onto goggles or from low-quality public terminals that provide only a grainy black-and-white appearance. Within the Metaverse, users appear as user-controlled avatars. In addition, system daemons may appear to maintain the Metaverse. Interestingly, the biological body of human users may be actually affected in that they suffer brain damage in the real world by merely looking at a bitmap image in a certain datafile named Snow Crash, as explained in more detail in the following.

2.1.1. Avatars and Daemons

Like any place in reality, the virtual Metaverse is subject to development. Developers can build buildings, parks, signs, as well as things that do not exist in reality, since the rules of three-dimensional spacetime can be ignored in the Metaverse. All these things are pieces of software, made available to the public over the worldwide fiber-optics network. They are graphic representations – the user interfaces – of a myriad different pieces of software that have been engineered by major corporations. In order to place these things in the Metaverse, they have had to get approval from the Global Multimedia Protocol Group (GMPG), have had to buy frontage, get zoning approval, obtain permits, bribe inspectors, etc. The money corporations pay to build things all goes into a trust fund owned and operated by the GMPG, which pays for developing and expanding the machinery that enables the Metaverse to exist.

The main character in Snow Crash is a freelance hacker named Hiro Protagonist, who is the top-ranked swordsman of all time in the Metaverse, but has to work as a pizza delivery driver to make ends meet in reality. Hiro and some of his buddies pooled their money and bought one of the first development licenses and created a little neighborhood of hackers in the Metaverse. By getting in on it early, Hiro's buddies got a head start on the whole business. Some of them even got rich off of it. That's why Hiro has a nice big house in the Metaverse, but has to share a 20-by-30 in reality. Real estate acumen does not always extend across universes.

Hiro wrote the sword-fighting algorithms, code that was later picked up and adopted by the entire Metaverse. He discovered that there was no good way to handle the aftermath of a sword fight since avatars are not supposed to die. So the first thing that happens, when someone loses a sword fight, is that his computer gets disconnected from the global network that is the Metaverse. It is the closest simulation of death that the Metaverse can offer. The user finds that he can't get back into the Metaverse for a few minutes. He can't log back on. This is because his avatar is still in the Metaverse and it's a rule that your avatar can't exist in two places at once. So the user can't get back in until his hacked-up avatar has been disposed of by graveyard daemons, a new Metaverse feature that Hiro had to invent, who emerge from invisible secret trapdoors climbing up out of hidden tunnels of the netherworld. The tunnel system is accessible only to the graveyard daemons. They will take the avatar to the pyre, an eternal, underground bonfire beneath the Metaverse, and burn it. As soon as the flames consume the avatar, it will vanish from the Metaverse and then its owner will be able to sign on as usual, creating a new avatar to run around in.

2.1.2. Hypercard and Language of Nature

One can't sell drugs in the Metaverse, because one can't get high by looking at something. But Snow Crash is computer lingo. It means a system crash – a bug – at such a fundamental level that it not only turns the VR goggles into a gyrating blizzard but also damages the brain. More specifically, Snow Crash is a *hypercard* that looks much like a business card. A hypercard can carry a virtually infinite amount of information or, more likely, a wide variety of nasty computer viruses. In fact, this Snow Crash thing is not only a *virus* but also a *drug* and, somewhat surprisingly, even a *religion*.

All people have religions. It's like we have religion receptors built into our brain cells, or something, and we'll latch onto anything that'll fill that niche for us. Religion used to be essentially viral – a piece of information that replicated inside the human mind, jumping from one person to the next. That's the way it used to be, and unfortunately, that's the way it's headed right now. But there have been several efforts to deliver us from the hands of primitive, irrational religion. The first was made by someone named Enki about 4000 years ago, followed by others such as Hebrew scholars in the eighth century BC. Another attempt was made by Jesus – that one was hijacked by viral influences within 50 days of his death. The virus was suppressed by the Catholic Church, but we're in the middle of a big epidemic, known as Pentecostalism, that started in Kansas in 1900 and has been gathering momentum ever since.[1]

Pentecostal Christians speak in tongues. The technical term is "glosso-lalia," which is a neurological phenomenon that is merely exploited in religious rituals. Pagan Greeks did it – Plato called it "theomania." The Oriental cults of the Roman Empire did it, as well as many other ancient and aboriginal cultures. For instance, the Tungus tribesmen of Siberia say that when the shaman goes into his trance and raves incoherent syllables, he learns the entire *language of Nature*. (Shamans in many cultures speak a sacred language, usually unintelligible to others, while in trance. In anthropology, "sacred language" refers to a unique shamanic language employed to speak with helping spirit[s].) The Sukuma people of Africa say that the language is "kinaturu," the tongue of the ancestors of all magicians, who are thought to have descended from one particular tribe. If mystical experiences are ruled out, then it seems that glossolalia comes from structures buried deep within the brain, common to all people.

2.1.3. Sumerian Creation Myth: The Nam-Shub of Enki – A Neurolinguistic Virus

Computers rely on the one and the zero to represent all things. This distinction between something and nothing – this pivotal separation between being and nonbeing – is quite fundamental and underlies many creation myths such as the *Sumerian creation myth* found on a tablet in Nippur, an ancient Mesopotamian city founded in approximately 5000 BC.[2] Sumerian was used

---

[1]According to BBC, Pentecostalism began among poor and disadvantaged people in the United States at the start of the twentieth century. During the last three decades of the twentieth century Pentecostalism grew very strongly, especially in the developing world where it poses a serious challenge to other, more established, denominations. Its growth is partly rooted in anger at widespread poverty, injustice, and inequality. Pentecostalism is not a church in itself, but a movement that includes many different churches. It is also a movement of renewal or revival within other denominations. Pentecostalism is often said to be rooted in experience rather than theology. Pentecostalism gets its name from the day of Pentecost (from the Greek *pentekostos*, meaning fiftieth), when, according to the Bible, the Holy Spirit descended on Jesus' disciples, leading them to speak in many languages, often referred to as "speaking in tongues."

[2]Sumer, or the "land of civilized kings," flourished in Mesopotamia, now modern-day Iraq, around 4500 BC. Sumerians created an advanced civilization with its own system of elaborate language and writing, architecture and arts, astronomy, and mathematics. According to ancient texts, each Sumerian city was guarded by its own god; and while humans and gods used to live together, the humans were servants to the gods. Sumerian mythology claims that, in the beginning, human-like

in Mesopotamia until roughly 2000 BC. The oldest of all written languages. Sumerian incantations demonstrate an intimate connection between the religious, the magical, and the esthetic so complete that any attempt to pull one away from the other will distort the whole. Nowadays, people don't believe in these kinds of things. Except in the Metaverse, that is, where magic is possible. The Metaverse in its entirety could be considered a single vast *nam-shub*, enacting itself on the Metaverse's underlying fiber-optics network. A nam-shub is a word from Sumerian and denotes a speech with magical force. The closest English equivalent word would be incantation (or magic spell).

At one point, everyone spoke Sumerian. Then nobody did. It just vanished. And there's no genocide to explain how the disappearance of the Sumerian language happened. The only thing that could explain that is some phenomenon that moved through the population, altering their minds in such a way that they couldn't process the Sumerian language anymore, resulting in the vast number of human languages consistent with the Tower of Babel story in the Bible. Kind of in the same way that a virus moves from one computer to another, damaging each computer in the same way, coiled like a serpent around the human brainstem. In fact, this phenomenon was a *neurolinguistic virus* that uses the human brain as a host. It was the nam-shub invented by a Sumerian god named Enki (briefly mentioned in Section 2.1.2), who spread it throughout Sumer using tablets made out of clay. More specifically, Enki was a neurolinguistic hacker, who was capable of programming other people's minds with verbal streams of data known as nam-shub. The neurolinguistic virus was able to transmute itself and actually alter the DNA of brain cells into a set of behaviors. Sumerian was a language ideally suited to the creation and prop-agation of neurolinguistic viruses. That a neurolinguistic virus, once released into Sumer, would spread rapidly and virulently, until it had infected everyone.

Early linguists, as well as Kabbalists, believed in a fictional language called the tongue of Eden, the language of Adam. It enabled all men to understand each other, to communicate without misunderstanding. It was the language of

---

gods ruled over Earth. When they came to the Earth, there was much work to be done and these gods toiled the soil, digging to make it habitable and mining its minerals. Anu, the god of gods, agreed that their labor was too great. His son Enki proposed to create man to bear the labor. A god was put to death, and his body and blood was mixed with clay. (Note that the word *human* comes from the Latin word *humus*, the organic component of soil, formed by the decomposition of leaves and other plant material by soil microorganisms.) From that material the first human being was created, in likeness to the gods. This first man was created in Eden, a Sumerian word which means "flat terrain." Eden is the garden of the gods and is located somewhere in Mesopotamia between the Tigris and Euphrates rivers. Opinions vary on the similarities between this creation story and the biblical story of Adam and Eve in Eden. Source: Sumerian Creation Myth: The origins of human beings according to ancient Sumerian texts. Read more: https://www.dvusd.org/site/handlers/filedownload.ashx?moduleinstanceid=149414&dataid=127660&FileName=Sumerian%20creation%20Story.pdf

the Logos, the moment when God created the world by speaking a word. In the tongue of Eden, naming a thing was the same as creating it. The speech of men is connected with divine speech and all language whether heavenly or human derives from one source: the Divine Name. The practical kabbalists, the sorcerers bore the title Ba'al Shem, meaning "master of the divine name."

Computers speak machine language. It's written in ones and zeroes – binary code. At the lowest level, all computers are programmed with strings of ones and zeroes. When you program in machine language, you are controlling the computer at its brainstem, the root of its existence. It's the tongue of Eden. But it's very difficult to work in machine language because you go crazy after a while, working at such a minute level. So a whole Babel of computer languages has been created for programmers: FORTRAN, BASIC, COBOL, LISP, Pascal, C, PROLOG, FORTH. You talk to the computer in one of these languages, and a piece of software called a compiler converts it into machine language.

The Snow Crash virus that eats through the brain is a string of binary information, shone into the face of humans in the form of a bitmap – a series of white and black pixels, where white represents zero and black represents one. The avatar who tried to infect Hiro got away, but left his bitmap behind. Hiro wrote a few simple programs that enable him to manipulate the contents of the bitmap without ever seeing it. The bitmap, like any other visible thing in the Metaverse, is a piece of software. It contains some code that describes what it looks like, so that your computer will know how to draw it, and some routines that govern it. And it contains, somewhere inside of itself, a resource, a hunk of data, the digital version of the Snow Crash virus.

We've got two kinds of language in our heads. The kind we're using now is acquired. It patterns our brains as we're learning it. But there's also a tongue that's based in the deep structures of the brain, that everyone shares. These structures consist of basic neural circuits that have to exist in order to allow our brains to acquire higher languages. We can access those areas of the brain under the right conditions. Glossolalia – speaking in tongues – is the *output* side of it, where the deep linguistic structures hook into our tongues and speak, bypassing all the higher, acquired languages. Everyone's known that for some time.

There's an *input* side, too. It works in reverse. Under the right conditions, your ears – or eyes – can tie into the deep structures, bypassing the higher language functions. Which is to say, someone who knows the right words can speak words, or show you visual symbols, that go past all your defenses and sink right into your brainstem. Like a hacker who breaks into a computer system, bypasses all the security precautions, and plugs himself into the core, enabling him to exert absolute control over the machine. In the same sense, once a neurolinguistic hacker plugs into the deep structures of our brain, we can't get him out – because we can't even control our own brain at such a basic level.

The Sumerian word for "mind" or "wisdom" is identical to the word for "ear." That's all those people were: ears with bodies attached. Passive receivers of information. But Enki was different. He had the unusual ability to write new verbal programs for humans – he was a hacker. He was, actually, the first modern man, a fully conscious human being. At some point, Enki realized that Sumer was stuck in a rut. People were carrying out the same old rules all the time, not coming up with new ones, not thinking for themselves. He realized that in order for the human race to advance, they had to be delivered from the grip of the old civilization. So he created the nam-shub of Enki, a virus that went into the deep structures of the brain and reprogrammed it. Henceforth, no one could understand the Sumerian language, or any other deep structure-based language. Cut off from our common deep structures, we began to develop new languages that had nothing in common with each other.

The nam-shub of Enki was the beginnings of human consciousness. It was the beginning of rational religion, too, the first time that people began to think about abstract issues like God and Good and Evil. That's where the name Babel comes from. Literally it means "Gate of God." It was the gate that allowed God to reach the human race. Babel is a gateway in our minds, a gateway that was opened by the nam-shub of Enki that broke us free and gave us the ability to think, moved us from a materialistic world to a dualistic world – a binary world – with both a physical and a spiritual component.

2.1.4. Metaverse: At Once Brand New and Very Ancient

In the Metaverse, the key realization was that there's no difference between modern culture and Sumerian. It is at once brand new and very ancient, getting us to observe the birth of a new religion. We have a huge workforce that is illiterate or alliterate and relies on TV – which is sort of an oral tradition. And we have a small, extremely literate power elite – the people who go into the Metaverse, basically – who understand that information is power, and who control society because they have this semimystical ability to speak magic computer languages. Enki's neurolinguistic hacking was developed as a new powerful technology by the monopoly owner of the Metaverse's underlying fiber-optics network. He also wanted to spread a biological virus by vaccinating people – and there was more than just vaccine in those needles. He devised a means for extracting the virus from human blood serum and packaged it as a drug known as Snow Crash. Further, he has taken glossolalia and perfected it, turned it into a science. He can control people by grafting radio receivers into their skulls, broadcasting instructions directly into their brainstems. If one person in a hundred has a receiver, he can distribute the instructions to all others. They will act out the instructions as though they have been programmed to. He also has a digital *metavirus* in binary code that can infect computers, or hackers, via the optic nerve. When it is placed into a computer, it snow-crashes the computer. But it is much more devastating when it goes into the mind of a hacker, a person who has an understanding of binary code built into the deep structures of his brain. The binary metavirus will destroy the mind of a hacker.

There's no way to stop the binary virus. But there's an antidote to it. The nam-shub of Enki still exists. He gave a copy to his son, who passed it on. Enki went out of his way to plant a message that later generations of hackers were supposed to decode. It would jam the people's mother-tongue neurons and prevent them from being programmed.

Neal Stephenson's idea of the Metaverse, which we summarized earlier, is by now widespread in the computer-graphics community and is being implemented in a number of different ways. The word "Metaverse" is his invention, which he came up with when he decided that existing words such as virtual reality were simply too awkward to use. In thinking about how the Metaverse might be constructed, he was influenced by the Apple Human Interface Guidelines, which is a book that explains the philosophy behind the Macintosh. After the first publication of Snow Crash he learned about a virtual reality system called Habitat, developed by F. Randall Farmer and Chip Morningstar. The system runs on Commodore 64 computers, and though it has all but died out in the United States, is still popular in Japan. Habitat includes many of the basic features of the Metaverse as described in Snow Crash.

2.2. Early Deployments

Since the publication of Neal Stephenson's novel in 1992, the number of white papers and peer-reviewed papers on the Metaverse has been growing constantly. For an up-to-date comprehensive review and taxonomy of past and present Metaverse publications, we refer the interested reader to [3]. In the following, we highlight the different Metaverse components, applications, and open challenges, followed by a couple of recent illustrative examples of early Metaverse deployments from both industry and academia.

2.2.1. Metaverse Components, Applications, and Open Challenges

According to [3], the Metaverse will utilize the following hardware and software components:

- HMD
- Hand-based input device

- Non-hand-based input device
- Motion input device
- Scene and object recognition
- Sound and speech recognition
- Scene and object generation
- Sound and speech synthesis
- Motion rendering

Apart from hardware and software components, content is the fundamental component that maintains the Metaverse. It is used to provide an immersive experience through well-organized stories and user-created events. In content, story reality, immersive experience, and conceptual completeness are important.

Rather than listing events in the Metaverse, it is important to find hidden relationships based on causal relationships between events and themes and construct a scenario line based on them. Unlike a text-based scenario, the Metaverse is more complex because it has to be configured in multi-modal and embodied environments. Each entity and relationship is used to organize events, and events must be organically combined to form scenario lines. Scenario lines construct the overall structure and serve as an index linking each event.

In order to compose a scenario line, it is necessary to connect events composed of entities and their relationship using a graph model. When user behavior data in a scenario graph is accumulated over the lifetime of the avatar, it is extended to the concept of life logging. Scenarios expand by adding entities and linking the added entities with relation. Scenario lines form a skeleton and expand entities and links to events to create rich stories. Connections between events and other events are formed by relationships and are linked within a scenario. In an event-based extended scenario, as the scenario lengthens and inconsistencies between events may occur, it is necessary to verify periodically whether the scenario does not conflict in concept. By instantiating the scenario graph, it hierarchically enlarges and contracts to verify that each event is organically connected and there is no contradiction.

Metaverse applications are mostly aimed at games, research and marketing simulations, and some office applications. At present, games are the most common platform in the popularization of the Metaverse, including so-called serious games. For instance, immersive rehabilitation systems use virtual reality to capture real-time user movements in gamified environments and execute complex movements to encourage self-improvement and competition. It is interesting to note that the Metaverse is more suitable for simulating social problems than conventional surveys and role-plays. This is due to the fact that avatars can change skin color and gender as desired, thereby reducing preconceived notions about social discrimination.

According to [3], unlike earlier Metaverse approaches, e.g. Linden Lab's Second Life, the current Metaverse is based on the social value of Generation Z that online and offline selves are not different. The younger generation considers the social meaning of the virtual world as important as that of the real world, since they think that their identity in virtual space and reality is the same. Extended reality (XR) is the medium that connects avatars in the Metaverse and users in the real world. Originally, the term avatar means an alter ego that has descended to the earth, a fundamental being, e.g. God, that has changed its form to human. The avatar was used as a predefined exaggerated form in the virtual world rather than reflecting the real world. However, it gradually changed into an ideal form that projects the outward appearance and reflects the real ego now rather than ideal alter ego. As a result, there have been instances where people in the real world commit crimes or show unethical behavior, e.g. sexual harassment, based on their online anonymity in the Metaverse.

Ethics, privacy, and security are critical issues because the Metaverse collects data on behavior that is more detailed than user conversations and history. Further, surveillance actions, e.g. inappropriate chat-room actions, and organizations that play the same role as police and government are needed in the virtual world. The norms and restrictions of the Metaverse may differ from those in the real world because they have a post-nationalism mindset and more degrees of freedom. Most Metaverse users belong to Generation Z with various social ideas. It will be necessary to build the Metaverse with a worldview and ethical consciousness, in which various types of avatar can live, rather than a Metaverse as a physical space. Toward this end, cross-disciplinary research is necessary, involving cognitive science, social sciences, psychology, and economics. Importantly, the virtual environment, in which people live using avatars, differs from how society currently operates. Neuroscience and psychological approaches should be used to better understand humans and create and maintain a deeper Metaverse. Finally, the virtual currency of the Metaverse is different from the currency in the real world. It can become a new variable from the point of economics and eventually develop into a fused form [3].

2.2.2. Nissan's Invisible-to-Visible (I2V) Connected-Car Technology Concept

Figure 2.1 illustrates a very interesting example of early Metaverse deployments in industry, namely Nissan's *invisible-to-visible* (I2V) connected-car technology concept that helps drivers "see the invisible" [4]. By merging

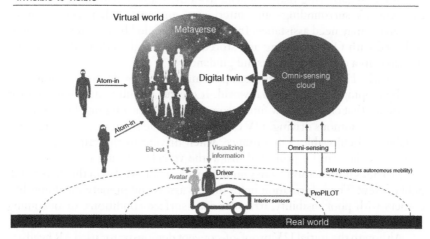

Figure 2.1   Nissan's invisible -to-visible (I2V) technology concept connects drivers and passengers to people in the Metaverse virtual world, who appear as three-dimensional AR avatars inside the car to provide company or assistance [4].

Source: Nissan Newsroom. Nissan unveils invisible-to-visible technology concept at CES: future connected-car technology merges real and virtual worlds to help drivers "See the Invisible," January 2019. (Online; accessed on 13 December 2022). Available: https://global.nissannews.com/ja-JP/releases/nissan-unveils-invisible-to-visible-technology-concept-at-ces.

information from sensors outside and inside the vehicle with data from the cloud, I2V enables the driver and passengers not only to track the vehicle's immediate surroundings but also to anticipate what's ahead, e.g. what's behind a building or around the corner. I2V can also connect drivers and passengers to people in the Metaverse. This makes it possible for family, friends, or others to appear inside the car as three-dimensional AR avatars to provide company or assistance. Guidance is given in an interactive, human-like way, such as avatars that appear inside the car. When visiting a new place, the system can search within the Metaverse for a knowledgeable local guide who can communicate with people in the vehicle in real time. The driver can also book a professional driver from the Metaverse to get personal instruction in real time. The professional driver appears as a projected avatar or as a virtual chase car in the driver's field of vision to demonstrate the best way to drive. Chasing a professional driver avatar can be used to improve driving skills. It can also be used by the onboard artificial intelligence (AI) system to provide a more efficient drive through local areas.

I2V is powered by Nissan's *Omni-Sensing* technology, a platform originally developed by the video gaming company Unity Technologies, which acts as a hub gathering real-time data from the traffic environment and from the vehicle's surroundings and interior to anticipate when people inside the vehicle may need assistance. Information provided by a local guide can be collected with Omni-Sensing and stored in the cloud so that others visiting the same area can access the useful guidance.

Clearly, I2V opens up endless opportunities for service and communication by tapping into the virtual world. It creates the ultimate connected-car experience that changes how future connected cars are integrated into society. During autonomous driving, I2V can make the time spent in a car more comfortable and enjoyable. For instance, when driving in the rain, the scenery of a sunny day can be projected inside the vehicle. During manual driving, I2V provides information from Omni-Sensing as an overlay in the driver's full field of view. The information helps drivers assess and prepare for things like corners with poor visibility, irregular road surface conditions, or oncoming traffic.

Although the initial I2V proof-of-concept demonstrator used AR headsets (i.e. wearables), Nissan envisions to turn the windshield of future self-driving cars into a portal to the virtual world, thus finally evolving from wearables to nearables.

2.2.3. Internet of No Things: Making the Internet Disappear and "See the Invisible"

Nissan's I2V connected-car technology concept can be expanded from the specific use-case scenario of future self-driving cars to the future Internet in general. Recall from Section 1.1 that the current gadgets-based Internet is anticipated to transit to a future Internet that is evolving from bearables, e.g. smartphone, toward wearables, e.g. AR headset, and nearables, thus giving rise to the Internet of No Things (see Figure 1.1). Similar to Nissan's I2V, in the Internet of No Things, all kinds of human-intended Internet services appear when needed and disappear when not needed.

The term Internet of No Things nicely resonates with Eric Schmidt's famous statement at the 2015 World Economic Forum that "the Internet will disappear" given that there will be so many things that we are wearing and interacting with that we won't even sense the Internet, even though it will be part of our presence all the time. Although at first this might sound a bit surprising, it is actually what profound technologies do in general. In "The Computer for the 21st Century," Mark Weiser argued that the most profound technologies are those that disappear. They weave themselves into the fabric of everyday life until they are indistinguishable from it [5].

In Chapter 5, we will build on Nissan's I2V technology concept and use it conjunction with other key enabling technologies to tie both online and offline worlds closer together in an Internet of No Things and make it "see the invisible" through the awareness of non-local events in space and time, including the creation of digital twin worlds that enable new superhuman capabilities and "sixth-sense" experiences.

2.3. Key Enabling Technologies

2.3.1. Metaverse Roadmap of 6G Communications and Networking Technologies

Recently, Tang et al. [6] outlined a roadmap to the Metaverse along with the requirements and development of key enabling 6G communications and networking technologies, as illustrated in Figure 2.2. In their presented framework, the authors discussed the fundamental technologies that need to be integrated in 6G in order to drive the implementation of the Metaverse, including intelligent sensing, digital twin, space-air-ground-sea integrated network (SAGSIN), edge computing, and blockchain, as discussed in more detail next.

- ▪ **Intelligent Sensing**: Sensing is the foundation of the Metaverse. Real-time information about the real world, including the surrounding environment, objects, and human beings, is fundamental to generate and update the virtual-reality space and is also essential to the interaction between the physical and virtual planes of the Metaverse. However, traditional sensing is not sufficient for the Metaverse to achieve the goal of a similar real world, as the sensing data is time-sensitive and easily hindered by network conditions. To enable an immersive experience of the Metaverse, real-time sensing is critical. 6G is required to support networked sensing for ubiquitous and high-accuracy sensing services, including a person's emotion in terms of outward expression inferred from audiovisual cues. Further, to measure inner feelings, some on-body sensors, including smart wearable and implanted devices, may be used to collect vital signs that change with our emotional states. A wide variety of wireless signals, such as Wireless Fidelity (WiFi), radio-frequency identification (RFID), Bluetooth, and Zigbee, have been used as powerful tools for sensing humans and the surrounding environment. After sensing, target activities can be detected and recognized accurately by extracting the unique features and putting them into machine learning models. Clearly, there is a potential to design integrated sensing, communications, and computing (ISACC) systems for achieving real-time networked

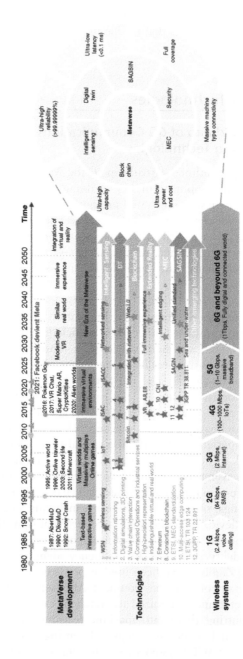

Figure 2.2 Roadmap to the Metaverse: requirements and development of key enabling 6G communications and networking technologies [6]. Source: F. Tang et al., 2022/IEEE © 2022 IEEE.

sensing. There is an urgent need for exploring new communications technologies to meet the data transmission requirements, introducing edge computing to process resource-intensive computation tasks, and employing AI algorithms to train models for activity recognition to improve computation efficiency. In addition, to guarantee scalability of collected information and improve its accuracy and coverage, new sensing paradigms such as participatory sensing and crowdsensing can be deployed to attract users and their contributed data.

- **Space-Air-Ground-Sea Integrated Network (SAGSIN):** Full coverage is essential to achieve global sensing and seamless connectivity, thereby providing unified experiences for users in the Metaverse. To achieve connectivity for intelligent sensing and communications in the Metaverse, in 6G, multi-dimensional networks called SAGSIN that deeply integrate space (e.g. satellites), air (e.g. balloons, drones, unmanned aerial vehicles [UAVs], etc.), ground (cellular, WiFi, wired networks), and sea layers were proposed to provide global seamless networking. With large coverage and dedicated bandwidth, UAVs can be used to achieve low-cost rural area coverage, while satellites can cover mountains and seas, thereby achieving global coverage. Furthermore, space-air networks can provide high-speed backhaul access for ground base stations through millimeter-wave and Terahertz (THz) communications, liberating wired backhaul infrastructures.

- **Digital Twin:** In the Metaverse, a comprehensive digital representation of the real world is created and maintained through digital twin technology. With digital twins, high-fidelity virtual models are created to truly reproduce the real world, such as geometric shapes, physical properties, behaviors, and rules, by using accurate modeling, communications, computation, and physical data processing technologies. Moreover, to provide quality-of-experience (QoE) for users, ultra-low latency in terms of communications and processing is required to achieve real-time information exchange of digital twins. Since the dynamic high-fidelity modeling relies on up-to-date accurate physical data and accurate modeling is the most important part of realizing a digital twin, it is required to ensure high transmission reliability. High privacy and security of data transmission between the physical and virtual planes of the Metaverse are also required. Therefore, the implementation of digital twins has strict requirements on 6G, including ultra-low latency, ultra-high reliability and throughput, and security. Moreover, to improve the robustness and reliability of twin data and expand the modeling dimension of virtual twins, it is necessary to fuse the collected data. Although effective AI algorithms can be applied for large data sets, there are still several challenges to train digital twin models for those large data sets. Federated AI can distribute different components of AI algorithms over the network. The data associated with digital twins and the corresponding AI algorithms may have to be cached in the cloud and/or several edge servers across the network. Thus, 6G is required to support seamless communications among distributed digital twins and computation of distributed AI algorithms for these digital twins, while ensuring data security and integrity.

- **Edge Computing:** Multi-access edge computing (MEC) is recognized as one of the key technologies in 5G/6G to achieve real-time mirroring between the real and virtual worlds. MEC is considered one of the key enabling technologies to support the Metaverse, which may be applied to efficiently process computation-intensive tasks and achieve real-time interactions in the Metaverse. However, there remain open issues to be addressed to support the Metaverse. Among others, efficient caching strategies should be designed to reduce the latency of content acquisition and mitigate the burden on the backhaul, thereby maximizing achievable quality-of-service (QoS) and minimizing storage cost. Also, new trust and authentication mechanisms should be developed to cater for the heterogeneity of MEC systems.

- **Blockchain:** Blockchain is considered a key enabling technology for the Metaverse. Decentralized and low-cost dynamic network management can be realized by blockchain. The improved interoperability between different systems of the Metaverse can be achieved by adopting a unified authentication system via blockchain. Further, reliability can be improved by distributing data across various network nodes. With improved interoperability, reliability, privacy, security, and scalability, blockchain is a key enabling technology for the Metaverse to achieve community integration, efficient resource management, and improved safety of both equipment and users.

2.3.2. CPSS-Based Sustainable and Smart Societies: Integrated Human, Artificial, Natural, and Organizational Intelligence

In Section 1.3, we have introduced the term cyber-physical-social system (CPSS) in our big-picture Metaverse discussion of narratives for 2030 and beyond. Recently, Fei-Yue Wang, who originally coined the term CPSS back in 2010, elaborated on the central role of CPSS in the emerging Metaverse.

In [7], he defines CPSS as the abstract and scientific name for metaverses, a term or source for inspiring us with more wild and innovative imaginations for a better world of *humanity* and *sustainability*. He explains that the concept of CPSS was inspired by Karl Popper's three-world philosophy on reality, i.e. physical, mental, and artificial worlds, supported by and operated in two spaces, the physical space and cyberspace. Two essential but contradictory kinds of processes and objectives in complexity science are unified and accomplished in CPSS, *emergence* in cyberspace and *convergence* in physical space for decision making and problem solving, and vice versa for problem identification and decision initialization. He argues that a concrete Metaverse is just a specific realization of CPSS.

In [7], Fei-Yue Wang outlines a journey toward the so-called TRUE decentralized autonomous organization (DAO) system for intelligent systems-based parallel intelligence with the help of digital twins, metaverses, Web3, and blockchain technology. He argues for a Hanoi approach, i.e. integrated human, artificial, natural, and organizational intelligence for achieving knowledge automation for *sustainable and smart societies*. Recent advances in blockchain, robotics, and AI, all in CPSS, have brought us the high hope and great potentials and possibilities we see in metaverses. Importantly, he argues, that for this, we need to rethink about the field of cognitive science to *bridge our cognitive gap* between the physical and mental worlds (to be further explored in Chapter 9). Parallel intelligence generates big data to fill up the gap and builds end-to-end bridges to connect different parts of the two worlds, from physical to mental and vice versa. Parallel intelligence provides an effective mechanism of making small data into big data, and then refining big data into deep intelligence for specific tasks, as demonstrated by deep learning and AlphaGo-like techniques.

Importantly, in [7], he outlines a new philosophy for intelligence of *Being, Becoming, and Believing*. According to Wang, the success of parallel intelligence and CPSS needs the support of new technologies developed in blockchains, DAOs, metaverses, and Web3. Specifically, the integration of blockchains, smart contracts, along with DAOs (DAOs type I) and their distributed autonomous operations (DAOs type II) is a *TRUE DAO*, i.e. a real journey leading to intelligent systems for smart societies. The term Dao or Tao in Chinese means journey or meta, and is actually the core concept of Chinese philosophy. In the West, the ancient root of metaverses comes from Greek philosophical thoughts on the fundamental nature of reality, the first principles of being in space and time, as described in Aristotle's *Metaphysics*. In addition, Wang postulates that we need to expand our philosophy of Being and Becoming to include a new "B," i.e. Believing. He argues that a new philosophy for intelligent humanity and smart societies with new IT, intelligent technology, of Being, Becoming, and Believing, by parallel intelligences, DAOs and CPSS, is in the urgent need. We must live with three Bs: Being for the physical world, Becoming of the mental world, and Believing in the artificial world.

Wang concludes that the new philosophy of intelligence will transform our worlds into *"6S" societies with "6I"*:

- Safe in the physical world
- Secure in the cyberworld
- Sustainable in the ecological world
- Sensitive to individual needs
- Serves for all and
- Smart in all

with

- Cognitive intelligence and parallel intelligence for intelligent science and technology
- Crypto intelligence and federated intelligence for intelligent operations and management
- Social intelligence and ecological intelligence for smart development and sustainability.

To this end, according to Wang, we must transform the tower of Hanoi, the toy problem used in the early studies of robotics and AI, into the TRUE tower of *Hanoi: Human, artificial, natural, and organizational intelligence.*

2.3.3. NFTs: Tulip Mania or Digital Renaissance?

Recently, Ross et al. [8] have published an interesting conference proceedings paper titled "NFTs: Tulip Mania or Digital Renaissance?," in which they make the observation that galleries, libraries, archives, and museums have begun to sell NFTs of works from their collections. For instance, the British Museum is one of several institutions that has recently sold NFTs, once again seeing them become stores of treasure, or treasuries, as they were in medieval times. Indeed, some argue that these institutions are sitting on a treasure trove of "NFT-able" digital assets, because of their large collections of archival documents, art, culture, pop, and design. To some, NFTs represent an innovative democratization of works and a means of generating new sources of revenue for cash-strapped institutions, especially during the global pandemic. Others see NFTs as crass commercialization and hype culture at its worst. Clearly, many open questions exist about whether NFTs are beneficial or harmful from financial, regulatory, and environmental perspectives. In their paper, Ross et al. [8] aimed at unpacking what NFTs are within the context of the emerging token economy and exploring open research challenges for the creation (i.e. minting) and purchase of NFTs, emphasizing NFTs' record-keeping abilities while also highlighting their inherent vulnerabilities. In the following, we briefly summarize their major findings.

2.3.3.1. What Is an NFT?

Normal tokens are also called *fungible tokens*. They are akin to currency: a dollar is a dollar is a dollar as the saying goes. Unlike an NFT, a fungible token is interchangeable (i.e. non-unique) and holds the same face value as other tokens of the same type. Fungible tokens can also be subdivided into smaller units of value.

In contrast, an NFT is not intended to be divisible, since the value of such a token is typically based upon the entirety of an asset. For instance, traditionally an artwork would not be the same artwork if it were to be divided into smaller parts.[3] An NFT is a unique, digital object certified on a blockchain or distributed ledger, typically Ethereum. Following Ethereum's creation in July 2015, the term NFT was coined in the Ethereum Request for Comments ERC-721 Ethereum Standard, published on 24 January 2018, after community consensus to present a diverse universe of assets, including physical property, virtual collectibles, as well as negative value assets such as loans or responsibilities. The emphasis is on tradeable assets: digital, physical, or both.

Although other blockchain platforms beside Ethereum can support NFTs, as of October 2021, only the Singaporean-centric open source blockchain TRON, founded by Justin Sun, CEO of BitTorrent, has created its own non-Ethereum NFT standard. Launched on 24 December 2020, TRON Request for Comments TRC-721 was created to mirror Ethereum token standard ERC-721.

2.3.3.2. How Do NFTs Work?

The intention of ERC-721 is to provide a standard application programming interface (API) for managing NFTs within the existing Ethereum smart contract structure, whether they be existing virtual collectibles, physical property, or negative value assets such as loans. Once an NFT has been created (i.e. minted) on, e.g. Ethereum, a process that involves a gas transaction fee, a record of its creation includes the owner's wallet address and a link to the NFT. Note that NFTs are always stored off-chain due to size and cost constraints, resulting in this link providing the only connection to the created object. As a result, although the record of the NFT is secure,

---

[3] Nevertheless, on 16 March 2021, the first auction of *fractionalized non-fungible tokens* (f-NFTs) occurred on fractional.art, whereby NFTs were broken down and sold as fractions of the original, purportedly to "democratize" NFTs and increase their liquidity. Thus, f-NFTs bridge between NFTs as collectibles and fungible tokens as interchangeable cryptocurrencies.

the NFT itself becomes dependent on the continuing existence of the webhost. Consequently, the neutrality and decentralization inherent to the blockchain ethos are lost, since these platforms come under the control of the elite few. Another consequence is fragility in the archival bond, the unique link between the transaction record on chain and the digital asset to which it refers. Users may check whether their NFTs are still accessible or if they have been lost due to a broken link, giving rise to the so-called *broken link problem.*

NFTs have been widely in circulation only recently and already there are many that have become lost. This is a fundamental flaw with existing NFT representations on blockchains. As an NFT file is typically too large, and therefore the gas cost to store on chain too expensive, a hash of the link to the NFT is provided in the smart contract instead. However, these off-chain storage solutions (i.e. external websites) are not subject to maintaining an immutable ledger; merely not renewing the web domain is enough for the NFT link to break and point to nowhere. Links also may be broken because the NFT is copied. In such cases, it becomes difficult to determine which copy is the original NFT and reassert ownership. While some NFT platforms attempt to address this issue by hosting their NFTs using inter-planetary file system (IPFS), a peer-to-peer (P2P) web-based file system that helps ensure that files are distributed across many hosts, this is not a universal practice. Further, IPFS relies on the popularity of a seeded file to continue access. If an NFT is no longer of interest to the community, it, too, suffers from the broken link problem and the file is effectively lost.

As the first blockchain to provide built-in smart contract functionality, Ethereum is the preferred platform for the buying and selling of NFTs. The largest and most established NFT trading platforms include Nifty Gateway, OpenSea, and SuperRare. They are all based on the Ethereum blockchain and as such require payment, both for purchase and transactional gas, in Ether (ETH). Thus, minting NFTs does not come without costs for institutions such as galleries, libraries, archives, and museums. Although traditional art auction houses such as Christie's and Sotheby's have facilitated payment of NFTs using either traditional fiat currencies (such as GBP or USD), in addition to more established cryptocurrencies (such as Bitcoin or ETH), all NFT platforms require an Ethereum-compatible cryptowallet to host the digital NFT.

2.4. Definition(s) of the Metaverse

In his long anticipated book "The Metaverse: And How It Will Revolutionize Everything," Canadian writer Matthew Ball argues that one of the most exciting aspects of the Metaverse is how poorly understood it is today [9].

2.4.1. Confusion and Uncertainty as Features of Disruption

Ball observes that for all the fascination with the Metaverse, the term has no consensus definition or consistent description. Most industry leaders define it in the manner that fits their own worldviews and/or the capabilities of their companies. For instance, Microsoft's CEO Satya Nadella has described the Metaverse as a platform that turns the entire world into an app canvas which could be augmented by cloud software and machine learning. No surprise, Microsoft already had a technology stack which was a natural fit for the not-quite-here Metaverse and spanned the company's operating system Windows, cloud computing offering Azure, communications platform Microsoft Teams, AR headset HoloLens, gaming platform Xbox, professional network LinkedIn, and Microsoft's own Metaverses including Minecraft, Microsoft Flight Simulator, and even the space-faring first-person shooter Halo. Conversely, Mark Zuckerberg's articulation focused on immersive virtual reality as well as social experiences that connect individuals who live far apart. Notably, Facebook's Oculus division is the market leader in VR in both unit sales and investment, while its social network is the largest and most used globally. Furthermore, the Washington Post characterized Epic's vision of the Metaverse as an expansive, digitized communal space where users can mingle freely with brands and one another in ways that permit self-expression and spark joy, a kind of online playground where users could join friends to play a multiplayer game like Epic's Fortnite one moment, watch a movie via Netflix the next and bring their friends to test drive a new car that's crafted exactly the same in the real world as it would be in this virtual one. It would not be the manicured, ad-laden news feed presented by platforms like Facebook. Facebook hasn't said whether or not the Metaverse can be privately operated, but the company does say that there can be only one Metaverse – just as there is "the Internet," not "an Internet" or "the Internets." In contrast, Microsoft and Roblox talk about "Metaverses."

According to Ball, the Metaverse is still only a theory. It is an intangible idea, not a touchable product. As a result, it's difficult to falsify any specific claim, and inevitable that the Metaverse is understood within the context of a given company's own capabilities and preferences. However, he notes that the sheer number of companies which see potential value in the Metaverse speaks to the size and diversity of the opportunity for widespread disruption. Far from disproving it, *confusion and uncertainty are features of disruption*. What makes technological transformation difficult to predict is the reality that it is caused not by any one invention, innovation, or individual, but instead requires many changes to come together. After a new technology is created, society and individual inventors respond to it, which leads to new behaviors and new products, which in turn lead to new use cases for the underlying technology, thereby inspiring additional behaviors and creations, giving rise to recursive innovation. The inability to precisely predict how we'll use the Metaverse, and how it will change our daily life, is not a flaw. Rather, it is a prerequisite for the Metaverse's disruptive force. The only way to prepare for what is coming is to focus on the specific technologies and features that together comprise it.

While there are competing definitions and a great deal of confusion, Ball believes that it is possible to offer a clear, comprehensive, and useful definition of the term, even at this early point in the history of the Metaverse. Ball provides the following useful working definition of the Metaverse:

> **A massively scaled and interoperable network of real-time rendered 3D virtual worlds that can be experienced synchronously and persistently by an effectively unlimited number of users with an individual sense of presence, and with continuity of data, such as identity, history, entitlements, objects, communications, and payments.**

Given the somewhat lengthy definition earlier, Ball adds that unpacking the etymology of the term Metaverse is helpful here. Neal Stephenson's neologism comes from the Greek prefix "meta" and the stem "verse," a backformation of the word "universe." For short, in English, "meta" roughly translates to "beyond" or "which transcends" the word that follows. In Chapter 10, we further elaborate on the term transcendence in the context of *transcendent experiences* as one of the most exciting opportunities brought forward by the future Metaverse. Moreover, Ball adds that many readers might be surprised that his definition is missing the terms *decentralization*, *Web3*, and *blockchain*. In recent years, these three words have become both ubiquitous and entangled with each other and with the term Metaverse. In fact, Ball states that the Metaverse and Web3 may arise in tandem. Large technological transitions often lead to *societal change* by tapping into widespread dissatisfaction with the present to pioneer a different future.

2.4.2. Blockchains, Cryptocurrencies, and Tokens: The First Digitally Native Payment Rail

Simply put, according to [9], the Metaverse is envisioned as a parallel plane for human leisure, labor, and existence more broadly. So it should come as no surprise that the extent to which the Metaverse succeeds will depend, in part, on whether it has a *thriving economy*. Yet we are not accustomed to thinking in these terms. While science fiction has predicted the Metaverse, one usually finds only glancing the references to a virtual world's internal economy in such stories. Toward this end, payment rails are an important requirement to achieve a flourishing and fully realized Metaverse. Ball argues that's why so many Metaverse-focused founders, investors, and analysts see *blockchains* and *cryptocurrencies* as the first digitally native payment rail and the solution to the problems plaguing the current virtual economy. Here, finally, we arrive at why there's such enthusiasm for blockchains.

Ball notes that some observers today believe that blockchain is structurally required for the Metaverse to become a reality, while others find that claim absurd. So why is a decentralized database or server architecture seen as the future? Ball argues that what matters is that blockchains are *programmable* payment rails. That is why many position them as the first digitally native payment rails, while contending PayPal, Venmo, WeChat, and others are little more than facsimiles of legacy ones. The greatest indicator of what blockchains might accomplish is what they have already achieved. For illustration, Ball highlights that in 2021, total transaction value exceeded $16 trillion – over five times as much than digital payment giants PayPal, Venmo, Shopify, and Stripe combined. In the fourth quarter, Ethereum (to be described in more detail shortly) processed more than Visa, the world's largest payment network and 12th-largest company by market capitalization. That this was possible without a central authority, managing partner, or even a headquarters – that it all happened via independent (and sometimes anonymous) contributors – is a marvel.

Not long after Bitcoin emerged, two early users – Vitalik Buterin and Gavin Wood – began developing a new blockchain, Ethereum, which they described as a "decentralized mining network and software development platform rolled into one." Like Bitcoin, Ethereum pays those operating its network through its own cryptocurrency, Ether. Importantly, however, Buterin and Wood also established a programming language, Solidity, that enabled developers to build their own permissionless and trustless applications called decentralized apps (DApps), which could also issue their own cryptocurrency-like *tokens* to contributors. In Chapters 6 and 7, we will discuss the emerging Web3 *token economy* and its closely related topics of *token engineering* and *tokenonomics* in technically greater detail.

Ethereum is a decentralized network that is programmed to automatically compensate its operators. These operators do not need to sign a contract to receive this compensation, nor worry about being paid, and while they compete with one another for compensation, this competition enhances the performance of the network, which in turn attracts more usage, thereby producing more transactions to manage. In addition, with Ethereum, anyone can program their own applications on top of this network, while also programming this application to compensate its contributors, and, if successful, providing value to those who operate the underlying network, too. All of this occurs without a single decision-maker or managing institution. In fact, there is and can be no such body. The decentralized governance approach does not prevent their underlying programming from being revised or improved. However, the *community* governs these changes and must therefore be convinced that any revisions are to their collective benefit. Ball mentions that most major public blockchains, in contrast to private blockchains, which are typically owned by a corporation, are decentralized and community-run.

The insatiable need for more computing resources and the long-held belief that realizing the Metaverse would require tapping into the billions of central processing units (CPUs) and graphics processing units (GPUs) that sit mostly unused at any given point in time. According to Ball, several blockchain-based startups are pursuing this – and they are succeeding. One, Otoy, created the Ethereum-based Render (RNDR) network and token so that those who needed extra GPU power could send their tasks to idle computers connected to the RNDR network, rather than to pricey cloud providers such as Amazon and Google. All of the negotiation and contracting between parties is handled within seconds by RNDR's protocol, neither side knows the identity or specifics of the task being performed, and all transactions occur using *RNDR tokens*. Another example given by Ball is Helium, which works through the use of $500 hotspot devices that allow their owner to securely rebroadcast their home Internet connection – and up to 200 times faster than a traditional home WiFi device. This Internet service can be used by anyone, from consumers to infrastructure, e.g. parking meter processing a credit card transaction. Those operating a Helium hotspot are compensated with Helium's *HNT token* in proportion to usage. As of 5 March 2022, Helium's network spanned more than 625,000 hot spots, up from fewer than 25,000 roughly a year earlier, distributed across nearly 50,000 cities in 165 countries. The total value of Helium's tokens exceeds $5 billion. Notably, the company was founded in 2013, but struggled to gain adoption until it pivoted from a traditional (i.e. unpaid) P2P model to one which offered contributors direct compensation via HNT tokens.

2.4.3. Social DAOs

Ball opines that the most disruptive aspect of digitally native programmable payment rails is how they enable greater independent collaboration and easier funding of new projects. For better understanding this in a broader context, he talks about a *vending machine*. The first of these devices actually emerged millennia ago (around AD 50) and allowed a consumer to insert a coin and receive holy water in return. By the late 1800s, these machines supported a wide variety of different purchases – not just a single item, such as water, but also gum, cigarettes, and postage stamps. No shopkeeper or lawyer managed the distribution of goods, nor accepted and validated payment, but the system worked through fixed rules: "If this, then this." Everyone trusted the system.

He continues that a blockchain vending machine would enable collabo-rators to write a smart contract, and after accepting each individual payment, the device would then automatically and incorruptibly deliver the appropri-ate amounts to the appropriate owner. Some envision smart contracts as the Metaverse-era version of the limited liability corporation (LLC) or nonprofit organization. A smart contract can be written and instantaneously funded, with no need for participants to sign documents, perform credit checks, con-firm payments or assign bank account access, hire lawyers, or even know the identities of the other participants. What's more, the smart contract trustlessly manages much of the administrative work for the organization on an ongoing basis, including the assignment of ownership rights, calculation of votes on bylaws, distribution of payments, and so on. These organizations are typically called *DAOs*.

According to Ball, many of the most expensive NFTs have been pur-chased not by individuals, but by DAOs comprising dozens (and is some cases, many thousands) of pseudonymous crypto users who could never have made the purchase on their own. Using the DAO's tokens, the collective can determine when these NFTs sell and at which minimum price, while also managing disbursements. The most notable example of such a DAO is the ConstitutionDAO, which was formed on 11 November 2021, to purchase one of the 13 surviving first editions of the United States Constitution, which was to be auctioned by Sotheby's on 18 November. Despite limited planning and no traditional bank account, the DAO was able to raise more than $47 million – far more than the $15 million–$20 million that Sotheby's estimated would be needed to win the auction. ConstitutionDAO ultimately lost to a private bidder, the billionaire hedge fund manager Ken Griffin, but Bloomberg, reporting on the effort, wrote that it "showed the power of the DAO … [DAOs have] the potential to change the way people buy things, build companies, share resources and run nonprofits."

More interestingly, Bell mentions that many *social DAOs* use smart con-tracts to issue tokens to individual members for their contributions, or to those who can't afford to join the collective but are deemed worthy by its members. He observes that some see social DAOs and tokens as a way to address tar-geted harassment and toxicity on large-scale online social networks. Imagine, for example, a model whereby Twitter users were awarded valuable Twitter tokens for reporting poor behavior, could earn more for reviewing previously reported tweets, and lost them if they violated the rules. At the same time, rather than rely on tips or posting promotional tweets on behalf of advertisers to generate income, super-users and influencers could be awarded tokens for hosting events. Ball notes that by the end of 2021, Kickstarter, Reddit, and Discord had all publicly described plans to shift to blockchain-based token models. Given their baked-in economic model, blockchains compensate those who contribute to their success or ongoing operations, rather than rely on altruism and empathy, as is the case with most open-source projects. Moreover, blockchain-based experiences seem, at least thus far, to promise developers far greater profits than closed platforms do.

2.4.4. Metaversal Existence: Life, Labor, and Leisure

According to Ball, ultimately, the Metaverse will be ushered in through experiences. Millions if not billions of users and dollars will be drawn to the new experiences and transformations that result. Notwithstanding, he warns that though these new experiences and resultant transformations are delivered, facilitated, or exacerbated by technology, the challenges we face in the mobile era are *human* and *societal* at their core. As more people, time, and spending go online, more of our problems go online, too.

He argues that the very idea of the Metaverse means that more of our lives, labor, leisure, time, spending, wealth, happiness, and relationships will go online. Actually, they will *exist* online, rather than just be put online like a Facebook post or Instagram upload. Many of the benefits of the Internet will grow as a result, but this fact will also exacerbate our great and unsolved *socio-technological challenges*. It is natural to worry about a future in which one goes outside and spends their existence strapped to a VR headset. Yet such fears tend to lack context. In the United States, for example, nearly 300 million people watch an average of five and a half hours of video per day (or 1.5 billion hours in total). Furthermore, we tend to watch video alone, on the couch or in bed, and none of it is social. As those in Hollywood often boast, this content is passively consumed. In industry jargon, it is lean-back entertainment. Ball advocates that *shifting any of the time passively consumed thru lean-back entertainment to social, interactive, and more engaged entertainment* is likely a positive outcome, not a negative one, even if we're all still indoors.

In Chapter 10, we will further elaborate on this important shift from passive to active experiences in the context of the so-called *experience machine*, a term coined by Harvard philosopher Robert Nozick in his 1975 national book award winning bestseller "Anarchy, State, and Utopia." The experience machine is an imaginative machine that produces favorable sensations by giving users whatever desirable experiences they might want. The experience machine gives users the choice between everyday reality and a presumably preferable simulated reality. Any future peak-experience machine should prioritize activities over passivities. Note that this would encourage people to actually plug into Robert Nozick's aforementioned experience machine, given that people want to do the actions, and not just have the experience of doing them, and have contact with a deeper, non-man-made reality, as we shall see in Chapter 10.

2.4.5. Gaming on the Blockchain

Finally, Ball addresses the important topic of gaming on the blockchain. According to Ball, regardless of one's long-term belief in NFTs, there are more interesting aspects of blockchain-based virtual worlds and communities. Tokens can be awarded for not only contributing time, delivering new users, data entry, intellectual property rights, capital, bandwidth, but also for good behavior such as community scores. These tokens can be provided with governance rights and, of course, may appreciate in value alongside the underlying project. Developers believe that this model can be used to reduce the need for investor funding, deepen their relationship with the community, and significantly increase engagement.

Ball argues that blockchain is not a technical requirement for these sorts of experiences, but many believe its trustless, permissionless, and frictionless structures make such experiences more likely to take off, thrive, and, most importantly, prove *sustainable*. Sustainability stems not just from increased user involvement in and ownership of an application, but from the ways blockchain discourages the application from betraying user trust and instead forces the application to earn it. In fact, in addition to operating DApps, Ball points out that blockchains can also be used to support the provision of compute-related *gaming* infrastructure.

More specifically, the scale and diversity of the crypto-gaming boom in 2021, matched with its relative infancy and enormous revenues per player, have led to a surge in development. Ball reports that nearly every talented game developer, with the exception of those already running world famous studios, was focused on building games on the blockchain. In total, blockchain-based games and gaming platforms received more than $4 billion in venture investment (total venture capital [VC] funding for blockchain companies and projects was roughly $30 billion; some speculate another $100 billion – $200 billion more has already been raised or earmarked by venture funds). The influx of talent, investment, and experimentation can quickly produce a virtuous cycle whereby more users set up a crypto wallet, play blockchain games, and buy NFTs, increasing the value and utility of all other blockchain products, which also attracts more developers, and in turn more users, and so on. Eventually, this leads us to a future in which a handful of exchangeable cryptocurrencies are used to power the economies of countless different games, replacing one where spending remains fragmented across Minecoins, V-Bucks, Robux, and countless other proprietary denominations.

Importantly, Ball argues that at enough scale, even the most successful game developers of the pre-blockchain era will find the technologies financially irresistible and competitively essential. The transition will be eased by the fact that they'll be opening up their economies and account systems to a system that is owned not by their platform competitors, but by the gaming community. In Chapter 10, we will expand on the rising importance of community experiences in our discussion of *team human* and its underlying transition from a narrow communications perspective to a broader *communitas* perspective, a well-known term in anthropology, thus deepening the intended human-centeredness of future 6G and Next G networks in support of the Metaverse.

2.5. Metaverse Economy and Community: Gamified Experiences

In one of the most recent books on the Metaverse, "Navigating the Metaverse: A Guide to Limitless Possibilities in a Web 3.0 World," [10] consolidated various perspectives to give a central definition of the Metaverse, or a *North Star*, as they call it:

> **The Metaverse represents the top-level hierarchy of persistent virtual spaces that may also interpolate in real life, so that social, commercial, and personal experiences emerge through Web3 technologies.**

According to [10], the Metaverse isn't meant to exist in a vacuum. Whether through VR, AR, or a smartphone, the Metaverse and the experiences inside it can connect to the real world. Further, *tokenomics* – to be explored in Chapter 7 – will be instrumental in providing a system of incentives which enables players to earn tokens in the Metaverse and purchase digital items such as NFTs, avatar clothing, quest items, and player equipment. As a result, these play-to-earn and play-to-own activities to earn tokens in the Metaverse create its own *Metaverse economy*.

In addition to incentivizing transactions, the Metaverse should also provide *gamified experiences*, which in turn lead to more transactions. Hackl et al. [10] observe that two seemingly unrelated things came together to change how the world viewed gaming. First, the Covid-19 pandemic decimated economies and put people out of work around the world. As a result, these people needed to think outside the box to pay bills and provide for their families. Second, blockchain games were cropping up on the fringes of mobile gaming. When the lockdown kept people from working, some found that they could earn tokens, enough to exchange for fiat money to get by in the dismal economic climate. Blockchain games and the Metaverse aren't just for killing or beating a game – it has the potential to redefine what it means to work for a living and put food on the table. If people can make a decent living while having fun, that's a strong alternative to manual labor or even white-collar jobs where uninspired employees wither away in cubicle farms from nine to five.

It's important to note that the Metaverse is more than Metaverse economy and blockchain games. There's a culture brewing here, one that sees the Metaverse as a platform to do more than what's humanly possible in the real world. Giving someone a product or service as a freebie is fine – they'll come collect. But giving users something to do with items and building an activity around NFTs creates higher retention and gives them a reason to come back for more. In fact, *community* is at the heart of the Metaverse economy. The Metaverse will be an amalgam of like-minded communities. Hackl et al. [10] argue that thinking of the users solely as consumers will sell you short of the Metaverse's true potential. When you find the right community, the authors' suggestion is to offer experiences that give assets a *purpose* by focusing on gamification, contextualization, and geolocation:

- Gamification encompasses the activity and story around NFTs.
- Contextualization emphasizes the importance of grounding your NFTs in the community and ecosystem.
- Geolocation means that NFTs and any component of a Metaverse project should evaluate ecosystems and specific locations in the real world to be in position to succeed.

Being the next step after the mobile Internet, Hackl et al. [10] conclude that the Metaverse is all about *evolution*, when we move away from our smartphones into wearables (in the 6G post-smartphone era) and when new AR hardware becomes more accessible to the masses.

In the remainder of this book, we will take a deeper dive into the various perspectives of the Metaverse mentioned earlier. In Chapter 6, we will delve into the emerging Web3 token economy, paying particular attention to the role of purpose-driven tokens. In the context of tokenomics, Chapter 7 highlights the beneficial impact of the blockchain-enabled trust game on human prosocial behavior. With regard to evolution, Chapter 9 explains why evolution makes us not only more social but also smarter. And, given the importance of community in the Metaverse, Chapter 10 touches on the possible transition from communication to communitas and its increasing degrees of perceived unity. Prior to that, however, the next chapter focuses on the important topic of experience innovation since experiences play such a central role in the Metaverse.

References

1. N. Stephenson. *Snow Crash*. Bantam Books, June 1992.
2. W. Gibson. *Neuromancer*. Ace, July 1984.
3. S.-M. Park and Y.-G. Kim. A Metaverse: taxonomy, components, applications, and open challenges. *IEEE Access*, 10:4209–4251, January 2022.
4. Nissan Newsroom. Nissan Unveils Invisible-to-Visible Technology Concept at CES: Future Connected-car Technology Merges Real and Virtual Worlds to Help Drivers 'See the Invisible', January 2019. [Online; Accessed on 2022-06-09]. Available: https://global.nissannews.com/ja-JP/releases/nissan-unveils-invisible-to-visible-technology-concept-at-ces.
5. M. Weiser. The computer for the 21st century. *Scientific American*, 265(3):94–104, September 1991.
6. F. Tang, X. Chen, M. Zhao, and N. Kato. The roadmap of communication and networking in 6G for the Metaverse. *IEEE Wireless Communications*, 1–15, June 2022. IEEE Xplore Early Access.
7. F.-Y. Wang. Parallel intelligence in Metaverses: Welcome to Hanoi! *IEEE Intelligent Systems*, 37(1):16–20, January/February 2022.
8. D. Ross, E. Cretu, and V. Lemieux. NFTs: Tulip Mania or Digital Renaissance? In *Proc., IEEE International Conference on Big Data*, pages 2262–2272, December 2021.
9. M. Ball. *The Metaverse: And How It Will Revolutionize Everything*. Liveright, July 2022.
10. C. Hackl, D. Lueth, T. Di Bartolo, and J. Arkontaky. *Navigating the Metaverse: A Guide to Limitless Possibilities in a Web 3.0 World*. Wiley, May 2022.

CHAPTER 3

The Multiverse: Infinite Possibility

"In terms of the Internet, nothing has happened yet. If you want a glimpse of what we humans do when the robots take our current jobs, look at experiences. Humans excel at creating and consuming experiences. This is no place for robots."

KEVIN KELLY
Founding Executive Editor of WIRED and Author of
"The Inevitable: Understanding the 12 Technological Forces
That Will Shape Our Future"
(1952–present)

3.1. Experience Innovation on the Digital Frontier

In Chapter 1, we have introduced the Multiverse as a powerful experience design canvas to create more sophisticated Extended Reality (XR) realms, including but not limited to Metaverse's Virtual Reality (VR) and Augmented Reality (AR). As shown in Figure 1.3, the Multiverse consists of three pairs of variables, each with two opposite physical/digital dimensions Space/No-Space, Time/No-Time, and Matter/No-Matter, which give rise to a total of eight possible realms. In this chapter, we describe the Multiverse's eight different realms in technically greater detail and further elaborate on its great potential for the design of advanced XR experiences in a comprehensive manner.

Pine and Korn [1] coined the term Multiverse in their book titled "Infinite Possibility: Creating Customer Value on the Digital Frontier." Digital technologies offer limitless opportunities, but real-world experiences have a richness that virtual ones do not. So how can you use the best of both

6G and Onward to Next G: The Road to the Multiverse, First Edition. Martin Maier.
© 2023 The Institute of Electrical and Electronics Engineers, Inc.
Published 2023 by John Wiley & Sons, Inc.

virtuality and reality? How do you make sense of such infinite possibility? What kinds of experiences can you create? To answer these and other questions, we review the major ideas and concepts of the Multiverse originally put forward by Pine and Korn [1].

What we desperately need in business today is *experience innovation*. We are now in an experience economy, where people desire experiences – memorable events that engage people in inherently personal ways, emotionally, physically, intellectually, and spiritually, more than the other economic offerings – and which have become the predominant economic offering.[1] The experience economy eclipsed the service economy that flowered in the latter half of the twentieth century, which in turn superseded the industrial revolution, which itself supplanted the agrarian economy (see also Figure 1.4). Experience innovations offer experiences, or even transformations built atop life-changing experiences, that engage customers and guide them in achieving their aspirations. These higher-order offerings create greater value for customers, generally have longer life spans as they prove more difficult for competitors to imitate, enable premium prices, and let companies capture more economic value.

The greatest source of offering experience innovation ever devised is digital technology. At the intersection of digital technology and experience innovation lies the *digital frontier*, where new customer value can be created by mining its rich veins of possibility. Experience innovations at the digital frontier enrich our lives by augmenting and thereby enhancing our reality, by engaging us through alternate views of reality that make us active participants in the world around us, by letting us play with time in ways not otherwise possible, by engrossing us in virtual worlds that enchant and capture our time, by letting us physically realize whatever we imagine, and by enabling virtual representations that mirror our reality to enlighten us from a new vantage point. Digital experience innovations can even give us a greater appreciation and desire for reality itself, whenever we take the time to unplug and just be. But by far the greatest value will come from those digital experience innovations that create *third spaces* that fuse the real and the virtual.

---

[1] Research shows that experiences make us happier than buying things. In the coming *Age of Experiences* [2], value will be determined more by how time is spent and less by how money is spent. More than in any of the other sectors of the economy (commodities, goods, and services), the currency that supports the experience economy is the *coin of time*. It is reasonable to predict that the demand for free leisure time will grow with the demand of experiences. In the early 1970s, famous futurist Alvin Toffler predicted that soon "we shall become the first culture in history to employ high technology to manufacture that most transient, yet lasting of products: the human experience." According to [2], the reason for the long delay has been the absence of some vital preconditions, prerequisites that have been met by three remarkable twenty-first century developments that have been slowly coming together in support of a new *eudaimonic technology*: (i) the dawn of the experience and transformation economies, (ii) the advent of the new sciences of happiness, and (iii) the coming of postmaterialist culture.

3.2. The Multiverse

Recall from Chapter 1 that the Multiverse offers a powerful experience design canvas to uncover hidden XR opportunities by fusing the real and the virtual, thereby creating cross-reality environments or so-called third spaces. Third spaces are created whenever one transverses the boundary between different XR realms within any given experience. Apart from conventional VR and AR, future XR technologies may realize novel, unprecedented types of reality. Thus, X may be rather viewed as a placeholder for future yet unforeseen developments on the digital frontier. The Multiverse may serve as an architecture for the design of such advanced XR experiences.

3.2.1. Realms of Experience

As shown in Figure 1.3, the Multiverse with its six variables delineates eight distinct *realms of experience*. It encompasses the multiple ways for *when* [Time ↔ No-Time] experiences happen, *where* [Space ↔ No-Space] they occur, and *what* [Matter ↔ No-Matter] they act on. Each and every combination of the six variables yields a distinct realm of experience. Table 3.1 provides an overview of the different possible combinations and the realms they create, spanning the entire reality-virtuality continuum. In the following, we describe each realm of experience in greater detail.

- **Reality**: Reality consists of the variables Time/Space/Matter. An equivalent way of looking at it is Actual/Real/Atoms (see Figure 1.3).

Table 3.1   Multiverse: variables and realms of experience.

Variables			Realms of experience
Time	Space	Matter	Reality
Time	Space	No-Matter	Augmented Reality
Time	No-Space	Matter	Physical Virtuality
Time	No-Space	No-Matter	Mirrored Virtuality
No-Time	Space	Matter	Warped Reality
No-Time	Space	No-Matter	Alternate Reality
No-Time	No-Space	Matter	Augmented Virtuality
No-Time	No-Space	No-Matter	Virtuality

Source: Pine and Korn [1].

We experience Reality as the realm of physical experiences through the age-old medium of real life. It is of course the realm with which we are most familiar.

■ **Virtuality**: Virtuality lies exactly opposite Reality in the realm of No-Time/No-Space/No-Matter, consisting of Autonomous/Virtual/Bits (shown in Figure 1.3). Virtuality is not subject to the physical laws of the real world. All Virtuality experiences sit atop Reality. Anyone having a Virtuality experience resides in some physical place, at a particular point in time, using a certain interaction devices such as VR goggles. Virtuality together with Reality anchor the Multiverse. While Reality is grounded firmly in our physical universe, Virtuality resides ethereally in the immaterial realm. The name of each of the other realms relates directly to the two anchors in that the names of each realm on the right half of Figure 1.3 denote their Reality-based nature, whereas the names of each realm on the left half of the Multiverse denote their Virtuality-based nature. These six other realms enhance, extend, or amend either our Reality- or Virtuality-based experiences. Therefore, they hold out greater possibility for value creation.

■ **Augmented Reality**: Of these, surely the most familiar is AR, characterized by the variables Time/Space/No-Matter. In the realm AR, digital technology is employed to enhance our experience of the physical world. The most obvious example is a global positioning system (GPS) navigation system, which enhances or augments your experience of the real world by making sense of it, providing directions to help you find your way.

■ **Augmented Virtuality**: If bits can augment Reality, then logically atoms should be able to augment Virtuality. This is exactly what happens in the opposite realm of augmented Virtuality (AV) characterized by the variables No-Time/No-Space/Matter, which is widely used for realizing digital twins. AV effectively flips a Virtuality experience from No-Matter to Matter, from Bits to Atoms. That means we're taking something material and tactile and using it to augment an otherwise virtual offering, resulting in an Autonomous/Virtual/Atoms experience. AV examples include haptic technology of sensor gloves that can manipulate virtual objects on screen. Another popular example is Nintendo's Wii, where, for the first time, players at home could get physically, materially engaged in computer games, removing the experience from one residing primarily between the fingers and the brain to one involving the whole body.

■ **Alternate Reality:** The shift from actual to autonomous events distinguishes Alternate Reality from the adjacent realm of AR, which both share the variables of digital substance and physical place. That means that if you can take the technology used to augment reality and then add a dimension of playing with time in some way, you can use that very same technology to alter people's view of the reality before them. Alternate Reality derives its name from alternate reality games (ARGs). Such games have become increasingly prominent in marketing circles as platforms for reaching the online gaming crowd. Alternate Reality comprises the variables No-Time/Space/No-Matter. In this realm of the Multiverse, Autonomous/Real/Bits experiences take games (and increasingly other activities) of the sort that normally play out online and take them from No-Space to Space, making the physical world a technologically infused playground of hyperlinked activity. Alternate Reality takes a virtual experience and shifts it over to the real world, making participants physically active in solving problems or discovering solutions. Its essence lies in constructing a digital experience and superimposing it onto a real place to create an alternate view of the physical reality. It blends the richness and sensations of the real with the power and complexity of the digital to challenge and involve participants in play environments where they learn almost every step of the way. With implications far beyond marketing, this realm starts with Reality and superimposes an alternate view on top of it.

- **Physical Virtuality**: Where Alternate Reality takes an otherwise virtual experience and plays it out in the real world, its opposite, Physical Virtuality, takes real-world objects (Atoms residing in Actual Time) and designs them virtually. Such a Time/No-Space/Matter experience occurs when virtually designed artifacts take material shape. 3D printing perhaps best captures the Actual/Virtual/Atoms nature of Physical Virtuality. With 3D printing, something designed virtually is printed, physical layer by physical layer in precise time sequence, to build up a material object from the experience.

- **Warped Reality**: The last realm on the Real side of the Space dimension, Warped Reality, consists of the variables No-Time/Space/Matter. As opposed to Augmented and Alternate Reality, this realm isn't about embracing digital technology or bringing virtual places into the real world. Rather, it takes an experience firmly grounded in Reality and shifts only one variable, moving the event from Actual to Autonomous Time. This realm of Autonomous/Real/Atoms is not infused with the digital technology of No-Matter, nor does it reside in the virtual arena of No-Space. It just requires the offering to play with or manipulate time in some way that makes it clearly distinct and different from normal experience. Such reality-based time travel happens whenever experiences simulate another time, e.g. Renaissance Fairs and living history museums, or transport us into the past or even into the future (albeit a fictional future), e.g. Star Trek conventions. So people leave Actual Time behind, departing Reality for another realm whose boundaries are that of imagination. A key aspect pertaining not only to Warped Reality but to each of the No-Time realms, is the concept of flow. The Hungarian psychologist Mihaly Csikszentmihalyi argued that we as human beings are most happy in flow, an *optimal experience* that evenly balances our skills with the challenges we face at a high level. Flow experiences exhilarate our emotions, excite our senses, energize our bodies, and elate our minds. And whenever the challenges we face slightly exceed our skills, they instigate a virtuous circle as we stretch and grow and enhance those skills, which incites us to seek out greater challenges, which yields further enhancements to our skills, and so on. Getting into flow may be the best way to embrace No-Time and thereby warp reality.

- **Mirrored Virtuality**: Finally, Mirrored Virtuality, characterized by the variables Time/No-Space/No-Matter, is the exact opposite to Warped Reality, where Virtuality is tied to Real Time. Inside Virtuality, operating via some sort of avatar, you generally remain free to do whatever you wish to do, whereas inside Mirrored Virtuality you inexorably remain tied into what is happening in the real world, in real time, moment by moment. This realm derives its name from the term *Mirror Worlds* coined by Yale computer scientist David Gelernter in 1992. In Mirror Worlds, virtual experiences tether themselves to what is going on in the real world, in real time. The best examples of such Actual/Virtual/Bits models can be found in Google Maps mashups such as Google Flu Trends (GFT), which beats the Center for Disease Control by analyzing searches for flu symptoms. The use of any sort of online tracking tool qualifies as Mirrored Virtuality since this realm offers a real-time view, a mirrored perspective, of what is going on out there, in the world.

Pine and Korn [1] emphasize that some of the greatest opportunities for creating customer value beyond the digital frontier will be discovered by operating on all the variables concurrently, effectively fusing realms into cohesive and compelling, rich and robust, individual and authentic *transversal experiences* never before envisioned, engendered, or encountered. The Multiverse pushes the frontier of experiences outward, opening up new galaxies begging for ambitious exploration, helping you explore strange new dimensions, seeking out new technologies and new experiences, and boldly going where no company has gone before. Further, they note that the three dimensions of the Multiverse do not stop here where they have drawn the boundaries of each realm. Expanding the eight aforementioned realms outward and encompassing ever more possibility are instrumental in creating deeper and more intense experiences through the innovations resulting from our imaginings.

3.2.2. Experience Design Canvas

Designers of experiences may use the so-called *experience design canvas* as a valuable tool to take full advantage of the six independent Multiverse variables. The Experience Design Canvas simply takes the six variables of the Multiverse and lays them out in a two-dimensional circle, as illustrated in Figure 3.1, so you easily can plot the intensity of the six variables within the experiences they together create. As our view shifts from the three-dimensional Multiverse (see Figure 1.3) to the two-dimensional experience design canvas, this new perspective deepens understanding and reveals opportunities. By plotting the variables of possibilities to depict them in a two-dimensional way lets us see the landscape of what's around us.

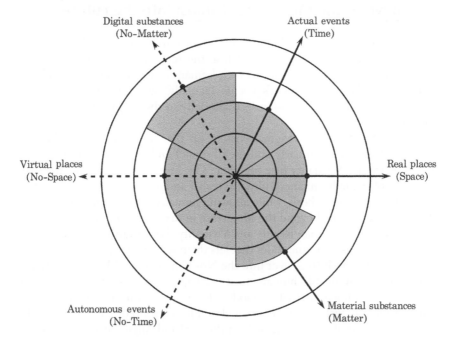

Figure 3.1   Experience design canvas: from three-dimensional Multiverse realms to two-dimensional variables.
Source: Adapted from Pine and Korn [1].

The experience design canvas serves as an ideation tool for discovering new value-creation experiences that extend beyond the strict confines of any one realm, often encompassing elements that have meaning in four, five, or even in all six variables. Any experience can be characterized by the six variables to reflect or capture its Actuality, Autonomy, Reality, Virtuality, Materiality, or Digitality – each generally at a different level of intensity. Once having depicted an existent experience variable by variable, you can then start thinking about and varying each of them individually and in their collective configuration to discover and depict just the right experience to meet your given needs, hopes, and desires.

Pine and Korn [1] recommend experience designers to focus their opportunity search on where the greatest value can be created – seek to solve customer problems, identify their latent ways that improve their lives, extend their current capabilities, and enable them to do things they otherwise are not able to do. As a result, they create experiences that eventually help enable the transformation of humans.

3.3. Beyond the Metaverse Origins: Into the Future

Recall from Section 3.1 that the experience innovation enabled by the Multiverse creates third spaces that fuse the real and the virtual by means of cross-reality environments, in which humans cross or transverse multiple realms and the environment's events (actually or autonomously), places (really or virtually), and substances (materially or digitally) within any given experience.

Neal Stephenson's original vision of the Metaverse, described in Chapter 2, provided an early and useful model and fictional conception for how Reality and Virtuality fuse together. The Metaverse represents the convergence of virtually enhanced physical reality and physically persistent virtual space. It is a fusion of both, while allowing humans to experience it as either. The Multiverse goes beyond the Metaverse in that it creates third spaces that involve realms other than Reality and Virtuality. For instance, an AR application could give users a time slider that lets them set surroundings into the past or future, thereby shifting over into the No-Time realm of AV. Alternatively, applications may enable humans to use a haptic device that provides force feedback operating on virtual 3D models. For instance, they may let you model with virtual clay just as if it were real clay. The addition of haptics causes the experience to reside simultaneously in Physical and Augmented Virtualities.[2]

---

[2] Recall from Section 2.1.3 the role clay plays in the Sumerian Creation Myth described in Neal Stephenson's novel, where a god was put to death and his body and blood was mixed with clay to create man.

3.3.1. Limitless Mind

Pine and Korn [1] conclude their book by asking the reader to reflect on what it means to be human. They argue that it has often been said that the mechanization that came with the industrial economy caused us to view ourselves in terms of machine metaphors – the brain as a determinant mechanism, for example. With the rise of the service economy and its dependence on information technology, computer metaphors abounded, with the brain as information processor. With the emergence of the experience economy and the Multiverse, they hope to see the concomitant rise of metaphors of imagination – the mind as limitless imagineer. We now truly are limited only by our imaginations, which stretch out to embrace infinite possibility.[3]

3.3.2. Realm of the Infinite: The Eternal

According to [1], we are not mechanisms or processors; we are ones who use the tremendous resources of digital technology to create what we imagine. Beyond the real and the virtual, beyond the universe of Time, Space, and Matter, and the Multiverse incorporating No-Time, No-Space, and No-Matter, lays another realm: *The Eternal*. In eternity, there exists the truly Infinite. The word "eternal" denotes, according to the Oxford English Dictionary, that which "always has existed and always will exist," that which is "not subject to time relations." In other words, eternity lies beyond both universe and Multiverse, both Reality and Virtuality, both actuality and possibility, both experience and imagination. Before any of these were, it is. In eternity exists the truly Infinite. And that existence speaks to the ultimate purpose for everything we do with the possibility that is set before us, through our acts and our selves as the means to reach beyond, to find the Infinite.

References

1. B. J. Pine II and K. C. Korn. *Infinite Possibility: Creating Customer Value on the Digital Frontier*. Berrett-Koehler Publishers, August 2011.
2. B. Hunnicutt. *The Age of Experience: Harnessing Happiness to Build a New Economy*. Temple University Press, February 2020.

---

[3] The perceptual ability of the limitless mind has been demonstrated and documented in numerous U.S. and international laboratories. For instance, the laboratory of Stanford Research Institute (SRI) in California has run a program of investigation for decades. In "Limitless Mind: A Guide to Remote Viewing and Transformation of Consciousness," Russell Targ, cofounder of the SRI program, argues that the brain can be described in part as a *quantum computer* and that consciousness emerges from quantum processes in the brain. More recently, the limitless mind was also investigated by Jo Boaler, a professor of education and equity at Stanford University, in her critically acclaimed book titled "Limitless Mind: Learn, Lead, and Live Without Barriers" (HarperOne, 2019).

CHAPTER 4

6G Vision and Next G Alliance Roadmap

"The best way to predict your future is to create it."

ABRAHAM LINCOLN
(1809–1865)

In Chapter 1, we have briefly reviewed some of the most intriguing 6G visions of the future life and digital society as of early 2021, when our prequel book was published. We saw that researchers focus not only on the technologies but they also expect the human transformation in the 6G era through unifying experiences across the physical, biological, and digital worlds. Furthermore, new themes are likely to emerge such as the creation of digital twin worlds that are a true representation of the physical and biological worlds at every spatial and time instant, unifying our experience across these physical, biological, and digital worlds. Digital twin worlds of both physical and biological entities will be an essential platform for the new digital services of the future. We also saw that digitalization will pave the way for the creation of new virtual worlds with digital representations of imaginary objects that can be blended with the digital twin world to various degrees to create a mixed-reality (MR), super-physical world, enabling new superhuman capabilities and augmenting human intelligence.

Recently, on 14 June 2022, the first IEEE Next G Summit took place.[1] Many of the presentations focused on the 6G vision and use cases. Given that the discussion on the 6G vision and use cases in International Telecommunication Union-Radiocommunication Sector (ITU-R) started in February 2021 and the final report will not be submitted until June 2023, the understanding of 6G use cases, enabling technologies, key performance indicators (KPIs),

---

[1] For further information, please visit: https://www.5gsummit.org/jhuapl/

6G and Onward to Next G: The Road to the Multiverse, First Edition. Martin Maier.
© 2023 The Institute of Electrical and Electronics Engineers, Inc.
Published 2023 by John Wiley & Sons, Inc.

network architecture, and spectrum aspects is still at a preliminary stage. Notwithstanding, one of the keynote speakers, Reinaldo Valenzuela from Nokia Bell Labs, showed that the defining new application for the 6G era will be immersive experience and extended reality (XR), followed by the digital–physical fusion, autonomous vehicles, and collaborative robots (co-bots) and artificial intelligence (AI) agents. Importantly, he pointed out that the Metaverse is more than a 3D immersive experience in that it enables not only the digital–physical fusion but also human augmentation. In fact, he makes the case that the suddenly unavoidable Metaverse presents a huge market opportunity of US$ 8–13 trillion by 2030 with a forecast total number of 5 billion Metaverse users worldwide. Furthermore, Alex Sprintson provided an overview of National Science Foundation (NSF)'s support for Next G research and clarified the difference between 6G and NSF's understanding of Next G, as explained in more detail shortly. For illustration, Figure 4.1 depicts NSF's near- to long-term Next G research objectives presented at the first IEEE Next G Summit. We observe from the figure that the Metaverse represents one of the important long-term Next G research objectives along with other important research topics such as quantum networks and holographic calls.

In this chapter, we review the recent progress toward realizing the 6G vision as well as the current state of the art of 6G research activities. Next, we

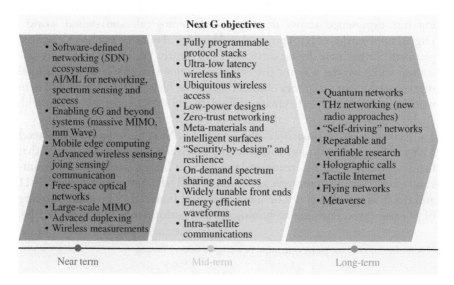

Figure 4.1   NSF's view on Next G research: near-, mid-, and long-term objectives. Source: Sprintson/© 2022 IEEE [1].

further elaborate on the aforementioned position of NSF on the difference between 6G and Next G research. Finally, we take a closer look at the recently published Next G Alliance roadmap to 6G.

4.1. 6G Vision: Recent Progress and State of the Art

4.1.1. 6G Paradigm Shifts

In Ericsson's recent 6G research outlook on connecting a cyber-physical world, the authors argue that there is no doubt that the ongoing societal transformation will give rise to challenges that 5G will be unable to meet [2]. Even with the built-in flexibility of 5G, we will see a need for expanding into new capabilities. This calls for further evolution – following the pull from society's needs and the push from more advanced technological tools becoming available – that must be addressed for the 6G era when it comes. Future networks will be a fundamental component for the functioning of virtually all parts of life, society, and industries, fulfilling the communication needs of humans as well as intelligent machines. According to Ericsson, four main drivers with corresponding challenges are emerging for the 6G era: (i) trustworthiness of the systems at the heart of society, (ii) sustainability through the efficiency of mobile technology, (iii) accelerated automation, and (iv) digitalization to simplify and improve people's lives.

More interestingly, Ericsson anticipates that addressing these challenges of the future also implies that there must be the following 6G paradigm shifts:

- **6G Paradigm Shift #1**: From secure communication to *trustworthy platforms*.
- **6G Paradigm Shift #2**: From data management to *data ownership*.
- **6G Paradigm Shift #3**: From energy efficiency to *sustainable transformation*.
- **6G Paradigm Shift #4**: From terrestrial 2D to global *3D connectivity*, including land, sea, and air areas.
- **6G Paradigm Shift #5**: From manually controlled to *learning networks* using intelligence to shift focus from instructing networks how to achieve goals to providing them with goals to achieve.
- **6G Paradigm Shift #6**: From predefined services to flexible *user-centricity* for a flexible network that should adapt to the needs of users and allow application to influence.
- **6G Paradigm Shift #7**: From physical and digital worlds to a *cyber-physical continuum* – the network platform should not only connect

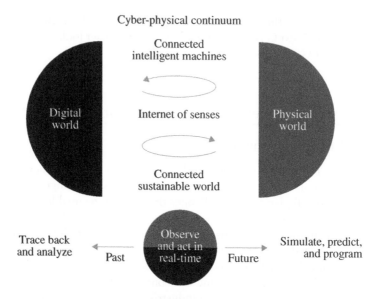

Figure 4.2 Ericsson's envisioned cyber-physical continuum for the 6G era. Source: Ericsson. 6G [2].

humans and machines but be able to fully merge realities to allow seamless interaction and immersive experiences.

■ **6G Paradigm Shift #8**: From data links to *services beyond communication* – expanding the role of networks to deliver services for a broad range of purposes as a versatile information platform.

Figure 4.2 depicts Ericsson's envisioned *cyber-physical continuum* for the 6G era to fully merge realities and allow immersive experiences. 6G makes it possible to move in a cyber-physical continuum, between the connected physical world of senses, actions, and experiences and its programmable digital representation. The network provides intelligence, limitless connectivity, and full synchronization of the physical and digital worlds. Vast amounts of sensors embedded in the physical world send data to update the digital representation in real time. Actuators in the real world carry out commands from intelligent agents in the digital world. It becomes possible to trace back and analyze events, observe, and act in real-time, as well as to simulate, predict, and program future actions. Similar to the Metaverse, the cyber-physical continuum provides a close link to reality, where digital objects are projected onto physical objects that are represented digitally, allowing them to seamlessly coexist as merged reality and enhance the real world.

6G should contribute to an efficient, human-friendly, sustainable society through ever-present intelligent communication. To serve as the future network platform for a vast range of new and evolving services, the capabilities of wireless access networks need to be enhanced and extended in various dimensions compared to networks of today. This includes classic capabilities, such as achievable data rates, latency, and system capacity, but also new capabilities, some of which may be more qualitative in nature [2].

4.1.2. Value-Creation-and-Capture Problem

In the 6G era, in which the convergence of wireless and Internet technologies will drive the digital economy, the focus on innovation will shift from individual technology or products toward innovation within platforms and ecosystems to have a transformational effect on society at large. To make sense of the future digital economy enabled by the convergence of wireless and Internet technologies, we need to envision 6G from technology, business, and societal perspectives in a multidisciplinary way. Unlike the existing 6G literature focusing on visions, use cases, and KPI requirements, Yrjölä et al. [3] extended the discussion to the *value-creation-and-capture problem* of technology innovators. The results of their technology value analysis show the emerging 6G business to be an oblique hybrid of vertical and horizontal business models and characterize the future 6G networks as a general-purpose technology (GPT). If 6G is to be a real GPT, it will be essential to shift the focus from separate protocol-layered technology innovations of focal firms, as in 5G, to dynamic multi-level innovation in platforms and ecosystems with novel business models to enable the creation and capture of value with 6G services and profiting from 6G innovations [3].

4.1.3. Perceptive Mobile Networks (PMNs)

Ubiquitous sensing and computing capabilities to interconnect millions of physical objects to the Internet are instrumental in realizing the 6G Internet of things (IoT). For a holistic survey on the convergence of 6G and IoT, we refer the interested reader to [4]. While wireless sensors are already ubiquitous, they are expected to be further integrated into wireless networks in the future. More precisely, sensing functionality could be a native capability of next-generation wireless networks.

Future dual-functional wireless networks supported by *integrated sensing and communications* (ISAC) technologies will not only serve as the foundation of the new air interface for 6G networks, but will also act as the bond to bridge the physical and cyber worlds, where everything is sensed, everything is connected, and everything is intelligent [5]. ISAC technologies offer an exciting opportunity of implementing sensing by utilizing wireless infrastructures, such that future networks will go beyond classical communication and provide ubiquitous sensing services to measure or even to image surrounding environments. This sensing functionality and the corresponding ability of the network to collect sensory data from the environment are seen as enablers for learning and building intelligence in the future smart world. Toward this end, ISAC attempts to merge sensing and communication into a single system, which previously competed over various types of resources. It pursues a deeper integration paradigm, where the two functionalities are no longer viewed as separate end-goals but are co-designed for mutual benefits via communication-assisted sensing and sensing-assisted communication. The role of existing cellular networks will shift to a ubiquitously deployed large-scale sensor network, which will trigger a variety of novel applications, e.g. human activity recognition, enhanced tracking, drone monitoring, spatial-aware computing, vehicle-to-everything (V2X), or simultaneous localization and mapping (SLAM). Importantly, equipped with sensing functionality, future mobile networks open their "eyes" and become so-called *perceptive mobile networks* (PMNs) [5].

4.1.4. Quantum-Enabled 6G Wireless Networks

Quantum information technology (QIT) has attracted extensive attention in recent years and a new wave of QIT research and development has arisen, although quantum mechanical principles have been known and studied for decades [6]. QIT is still evolving, but there is no doubt that QIT will enable novel ICT applications by leveraging quantum computing, quantum communications, and even quantum sensing. QIT will enable and boost future 6G systems from both communication and computing perspectives. For example, secure quantum communications such as quantum key distribution (QKD) can be leveraged to improve 6G security. Further, quantum computing can solve computationally difficult optimization problems in 6G such as wireless resource allocation to find optimal solutions. In addition, quantum machine learning (ML) can greatly promote ubiquitous wireless AI in 6G as well. It is also possible that quantum sensing can enable new types of haptic communications in 6G. Recent advances in photonic quantum computing mean that room-temperature-based quantum hardware is now within technological reach. In fact, photonic-based quantum hardware will most easily be able to integrate with the vast existing base of fiber-optical-based communication equipment leading to a potential realistic path toward 6G.

According to a recent survey on quantum-enabled 6G wireless networks by Wang and Rahman [6], QIT exploits principles which do not have counterparts in classical systems. These quantum mechanical principles are as follows:

- **Non-cloning theorem** states that it is impossible to duplicate an existing quantum state. This provides the foundation for information-theoretical security. For instance, if an eavesdropper tries to intercept a photon in the middle of the path from a sender to a receiver, it will be automatically detected by the receiver due to the non-cloning theorem. This phenomenon is leveraged in QKD protocols.
- **Superposition** is another unique quantum mechanical phenomenon describing quantum states. In classical systems, the state of the value of a classical bit is deterministic (i.e. 0 or 1). Conversely, a quantum state of a qubit is non-deterministic and can be a superposition of two (or more) basic states. Thanks to superposition, a qubit can concurrently represent two states and n qubits can denote 2^n potential states in total. As a result, computation on n qubits can be regarded as computation on 2^n states simultaneously. This is the origin of quantum computation speedup.
- **Entanglement** is an even stranger but extremely useful phenomenon, where quantum states of two (or more) qubits are maximally entangled. In other words, the quantum state of one entangled qubit is fully dependent on the state of any other entangled qubit, no matter how far apart the entangled qubits are physically located. Entangled qubits are the most crucial resources in QIT systems. They can enable new quantum procedures and applications such as quantum teleportation, which allows transmitting the state of a qubit from one quantum computer A to another quantum computer B without physically transmitting the qubit. This phenomenon occurs due to entanglement and is known as the principle of nonlocality, also referred to as *quantum-interconnectedness of all things* by quantum physicists such as David Bohm, which transcends spatial and temporal barriers. That's why Albert Einstein famously called quantum entanglement *"spooky action at a distance."*

4.1.5. From Softwarized to Blockchainized Mobile Networks

Along with various other candidate technologies, blockchain is envisioned to enable and enhance numerous key functionalities of 6G [7]. While smart contracts can softwarize all the terms and conditions of an agreement between various entities for the sake of access control, trust building, and elimination of third-party execution, the innate and pervasive integration of blockchain in the 6G ecosystem is conceptualized as *blockchainized* 6G [8]. Since the vision of blockchainized 6G goes well beyond the 5G case, it is essential to explore what more the blockchain technology can bring to the 6G realm. For instance, blockchain has immense potential to enable or enhance numerous key 6G functionalities such as automation of network management, dynamic spectrum management at THz communications, AI-powered edge computing, federated resource sharing in trustless environments, and security of ML and federated learning models.

Although 6G is in the infancy stage, Kalla et al. [8] categorize the different aspects of 6G into the following four emerging directions that anticipate significant advancements leveraging blockchain in accordance with the vision proposed by Europe's 6G flagship project *Hexa-X²*:

- **Trust-Based Secure Networks (TBSN)**: TBSN address trustworthiness by enabling 6G to promise trust, security, and privacy for connected digital services.
- **Harmonized Mobile Networks (HMN)**: HMN empower 6G to emerge as a network by aggregating resources such as communication, computing, sensing, and intelligence at different scales, ranging from an in-body environment to space.
- **Hyper Intelligent Networks (HIN)**: HIN realize connected intelligence and extreme experiences by ubiquitously integrating AI and ML in different segments of 6G mobile networks.
- **Resource-Efficient Networks (REN)**: REN aim at sustainability and global service coverage by realizing 6G as a highly optimized communication infrastructure that largely reduces the global ICT environmental footprint.

4.1.6. AI-Native 6G Networks

4.1.6.1. From 5G AI4Net to 6G Net4AI

Computing power plays a critical role in the development of AI. Tong and Li [9] argue that unfortunately, neither brute-force computing nor endless data is sustainable, and therefore identified nine AI challenges that should be addressed for 6G. Among others, reducing the computational complexity becomes the primary task for the future AI. Even if some techniques, such as pruning, low-dimensional compression, fewer quantization levels, or smaller deep neural networks (DNNs), can be applied to mitigate the issue, AI computing is still a long-term engineering problem in the following decades.

---

²The Hexa-X vision is to connect human, physical, and digital worlds with a fabric of 6G key enablers. For further information, please visit: https://hexa-x.eu

Distributed neural networks and distributed learning make it possible to share the computational load among different devices or neural networks. In general, the larger the memory capacity of a neural network is, the higher is its potential to improve communication efficiency. The memory capacity for a general DNN is still unknown. Another correlated question is the relationship between the memory capacity and the computational complexity. Is it linear, exponential, or some other relation?

Furthermore, Tong and Li [9] note that the big data required by training DNNs is often collected by wireless communications. Therefore, it becomes critical how to train DNNs efficiently and quickly to lower the requirement on big data and wireless communications. There have been some initial studies on exploiting the domain knowledge in the area of communications to reduce the requirement on training data, such as model-driven deep learning (DL) for communication systems. It is a tricky issue to trade off the domain knowledge usage, big data requirements, communication performance, and system complexity. Currently, AI models are trained under the special learning hypothesis premise that the environment (i.e. the statistics of the training data) is static, at least during the training period. However, real communication scenarios, especially for mobile communications, are constantly changing. The basic features of dynamic DL for wireless communications require extensive investigations. The challenge can be potentially addressed with the developments of accretionary learning and meta learning. The theory of accretionary learning for human beings was proposed by cognitive psychologists over 40 years ago. Accretionary learning can address the issue by extending DNNs through interaction of accretion, fine-tuning, and restructuring. Moreover, meta learning, also known as learning-to-learn, has become an active research area recently. With its development, dynamic learning for communication systems and networks could be achieved one day.

Another interesting AI challenge identified by Tong and Li [9] is wireless data-aided collective learning, where multiple servers collect data and train ML models collectively. Collective learning belongs to a general category of distributed learning, though with a special relation to communications. In the integrated architecture of AI and wireless communications, the primary problems are how to segment data and how to separate communication and computing. Specifically, how to segment a neural network and its tasks for collective learning is an unsolved problem, which is also a profound theoretical issue. Some related research problems are wireless communications-enabled federated learning, a type of distributed learning enabled by wireless communications. Different from collective learning, federated learning relies more on communications. Multiple clients/terminals jointly train the same model in federated learning, and there are only parameter exchanges among multiple terminals, rather than segmenting data, tasks, and neural networks, as done in collective learning. The information exchange between the terminal and edge neural networks depends on wireless communications.

Interestingly, Tong and Li [9] also clarified the following couple of concepts between AI and wireless networks:

- **AI for Communication Network (AI4Net)**: Since several years, we have been focusing on using AI to improve the intelligence and transmission performance and to reduce the complexity of communication systems, which is regarded as AI4Net. AI4Net mainly belongs to 5G and 5.5G.

- **Communication Network for AI (Net4AI)**: In the future, wireless communications will be an integrated part of AI for performance optimization, in addition to data collection and transmission, which is Net4AI. Net4AI is the mainstay for 6G and beyond.

4.1.6.2. Intelligence-Endogenous Network (IEN)

The emergence of new 6G services and applications like holographic communications and digital twins not only puts forward high requirements on traditional KPIs, such as delay, data rate, and reliability, but also places extremely high demands on network intelligence. At present, quite a few AI methods have been applied to network research, but most of them merely use ML algorithms to solve specific network problems. The current network intelligentization mechanisms based on rule-based algorithms are limited by rigid preset rules. As a result, it is difficult to adapt dynamically to constantly changing user needs and network environments. That is, under the present network intelligentization paradigms, networks do not have the capability of *self-evolution*. Adding "intelligence genes" into the network to form intelligence and self-evolution capabilities is an important way to resolve the aforementioned problems and give rise to so-called *intelligence-endogenous networks* (IENs) [10]. Note that an IEN with self-evolution capabilities is an interesting example of biologization, which we have introduced in Section 1.3.2.

According to [10], an IEN is centered on knowledge. It introduces AI technology into networks to characterize, construct, learn, apply, and update the multi-dimensional subjective and objective knowledge of the network. Based on the knowledge graph and stereo perception, decision making for dynamic adjustment of the network can be realized, so the network can change as it needs for whatever new services users want. An IEN is composed of a double-layer closed-loop structure that can realize self-evolution. More specifically, the inner layer represents the self-evolving core, also referred to as "knowledge brain," which takes the network knowledge graph as the core in order to generate endogenous intelligence. The network knowledge can be dynamically updated according to the changes in the actual environment and the evaluation of each network optimization effect. The self-evolving core affects the intelligence of the network in three aspects. First, through accurate analysis of service and user preferences, the IEN provides users with appropriate network services more intelligently. Second, the intelligentization of network resource management perceives and evaluates the impact of existing resource management strategies on the network operation, continues to improve these strategies based on the evaluation results, and distributes the improved strategies to various IEN agents. Third, the intelligentization of network infrastructure management uses the accumulated field experience and network knowledge path to provide upgrading, adding, and other improvement strategies for the network infrastructure.

Conversely, the outer layer is composed of a series of endogenous intelligence-oriented closed-loop network operation and management functions. The outer layer provides data and experience input to the self-evolving core and can realize the planning, deployment, operation, and adjustment of the network according to the knowledge provided by the self-evolving core, as well as realize the self-evolution of the network operation and management strategy [10].

4.1.6.3. Native Edge AI: Mimicking Nature Through Brain-Inspired Stigmergy

Recall from our discussion of the Internet of No Things in Section 5.1.1 that in contrast to previous generations, 6G will be transformative and will revolutionize the wireless evolution from "connected things" to "connected intelligence." Further, we saw that 6G will play a significant role in advancing Nikola Tesla's prophecy that "when wireless is perfectly applied, the whole Earth will be converted into a huge brain."

Recently, Letaief et al. [11] envisioned that AI will become a *native* tool to design disruptive wireless technologies for accelerating the design, standardization, and commercialization of 6G. In fact, the authors cited our work on the Internet of No Things as a prime example of advanced XR experiences and the Metaverse, which will go beyond the mobile Internet to support ubiquitous AI services. In particular, the United States, European Union, and China have funded 6G projects with a common goal of enabling connected intelligence. According to [11], however, state-of-the-art DL and big data analytics-based AI systems require tremendous computation and communication resources, causing significant latency, energy consumption, network congestion, and privacy leakage in both of the training and inference processes.

To improve the efficiency, effectiveness, privacy, and security of future 6G networks, Letaief et al. [11] provide their vision for scalable and trustworthy *edge AI* that represents a promising solution for connected intelligence by enabling data collection, processing, transmission, and consumption at the network edge. Specifically, by embedding the training capabilities across the network nodes, edge training is able to preserve privacy and confidentiality, achieve high security and fault-tolerance, as well as reduce network traffic congestion and energy consumption. Importantly, by directly executing the AI models at the network edge, edge inference can provide low-latency and high-reliability AI services by requiring less computation, communication, storage, and engineering resources.

At the downside, however, edge AI will cause task-oriented data traffic flows over wireless networks, for which disruptive wireless techniques, efficient resource allocation methods, and holistic system architectures need to be developed. To embrace the era of edge AI, wireless communication systems and edge AI algorithms need to be co-designed for seamlessly integrating communication, computation, and learning. In particular, differential privacy, Lagrange coded computing, security multi-party computation, quantum computing, blockchain, and distributed ledger technologies can be further leveraged to build trustworthy edge AI architectures. However, with limited storage, computation, and communication resources in wireless edge networks, a paradigm shift for wireless system design is required from data-oriented communication (i.e. maximizing communication rate or reliability on Shannon theory) to task-oriented communication (i.e. achieving fast and accurate intelligence distillation at the network edge) [11].

It is interesting to note that [11] observe that edge AI paves the way for network sensing and cooperative perception to understand the network environment. Importantly, to further imbue native intelligence, the authors note that *mimicking nature* for innovative edge AI empowered future networks can be envisioned. Specifically, brain-inspired *stigmergy*-based federated collective intelligence mechanisms hold promise to accomplish multi-agent tasks through simple *indirect communication*. In Chapter 8, we will further explore the benefits of mimicking nature and in particular stigmergy and indirect communication to maintain social cohesion by the coupling of environmental and social organization and evolution of social life toward a future *stigmergy enhanced Society 5.0*. Nature-inspired edge AI models and network architectures provide a strong evidence that one can establish an integrated data-driven and knowledge-guided framework to design and optimize 6G networks. Edge AI serves as a distributed neural network to imbue connected intelligence in 6G, thereby enabling intelligent and seamless interactions among the human world, physical world, and digital world, acting pretty much like the global mind or world brain introduced in Section 1.1.

4.1.7. THz Communications: An Old Problem Revisited

Terahertz (THz) band (0.1–10 THz) communications is envisioned as one of the key enabling technologies for 6G and beyond to support peak data rates of 1 Tbps. According to [12], the THz band reveals its potential as one of the key wireless technologies to fulfill the future demands for 6G wireless systems, thanks to its four strengths: (i) from tens and up to hundreds of GHz of contiguous bandwidth, (ii) picosecond-level symbol duration, (iii) integration of thousands of sub-millimeter-long antennas, and (iv) ease of coexistence with other regulated and standardized spectrum.

Ten years ago, the research on THz communications was mostly theoretical. Today, thanks to the advancements in device technologies, THz research is quickly transitioning from theory to practice, as witnessed by numerous existing THz experimental platforms as well as state-of-the-art simulation and emulation software, which serves as an intermediate step between pure theory and experimental research. Enabling 1 Tbps links requires not only the development of analog front-ends and antenna systems that can operate at higher frequencies and with larger bandwidths, but also the development of a digital signal processing (DSP) back-end that can generate at the transmitter and process at the receiver the information at such very high data rates. For the time being, all DSP solutions rely on electronic devices, mostly because of all the commercial computing cores are electronic systems. While there are early concepts of fully optical processors and even plasmonic processors, these are not likely to become a reality within 6G systems [12].

The road to close the THz gap and realize THz communications needs to be paved by the communication community jointly. In this direction, in 2019, the Federal Communication Commission (FCC) created a new category of experimental licenses and allocated over 20 GHz of unlicensed spectrum between 95 GHz and 3 THz to facilitate the testing of 6G and beyond technologies. The standardization activities have been on the way since several years. Recently, a new ITU-R report entitled "IMT Above 100 GHz" was started at the August 2021 meeting of ITU-R WP5D to study the technical feasibility of International Mobile Telecommunications (IMT) in bands above 100 GHz. Further, in July 2021, IEEE Communications Society's Radio Communications Committee Special Interest Group on THz communications has been established with the objectives and mission to advance the research and development of THz communications in 6G and beyond [12].

Despite the numerous technical achievements made in the last decade, there are several issues that need to be carefully addressed. Some of the most challenging issues that need to be overcome for future wireless systems at THz frequencies, including not only 6G networks but also short-range connectivity and fixed wireless links, were discussed in [13]. Several critical digital hardware limitations hinder achieving high data rates such as a terabit per second peak data rate envisioned in 6G. Specifically, the design of power-efficient analog-to-digital converters (ADCs) and digital-to-analog converters (DACs), power- and spectrum-efficient waveforms, line-of-sight (LOS) THz multiple-input and multiple-output (MIMO), as well as beam-forming transceivers at THz frequencies are among the most important tasks for future THz communications systems.

4.1.8. Comparison between NG-OANs and 6G RANs

It is important to mention that the evolution of mobile networks will also need new optical transport architectures, which are able to handle high-speed data flows on the order of hundreds of Gigabits per second and face new challenging requirements in terms of lowering costs, energy consumption, and footprint. The increase of processing capabilities of next-generation radio systems will create demand for new opto-electronic hardware archi-tectures and devices, including photonic interconnects, multi-chip modules, and co-packaged optics. Radio frequency (RF) and microwave electronics together with optical technologies will allow the realization of future gen-erations of integrated radio subsystems. Microwave photonics components will be key enablers for those systems to achieve photonic signal processing elements, devices for generating high-stability clock, carrier frequencies, and waveforms at very high frequencies. A key role will be played by inte-grated photonic technologies to realize the low-cost components for optical transport, interconnects, and optoelectronic devices for radio systems [14].

An insightful comparison between next-generation optical access networks (NG-OANs) and 6G radio access networks (RANs) on several critical metrics was conducted recently by Guo et al. [15]. NG-OANs will adopt next-generation passive optical networks (NG-PONs) to provide super-broadband services in support of super-broadband applications such as 3D holographic communication and cybertwins of the physical world. As the evolution of current PONs, NG-PONs are expected to provide higher data rates, larger amounts of wavelength channels, longer fiber ranges, and higher splitting ratios as well as broader functionalities. As two mainstream super-broadband access technologies, both NG-OANs and 6G RANs cannot meet the diverse requirements of future application scenarios. It should be better to select the more suitable one in different scenarios based on their respective pros and cons in Table 4.1. In fact, Guo et al. [15] propose to develop integrated fiber-wireless (FiWi) access networks, which leverage the complementary advantages of these two technologies, as a promising solution to implementing the future super-broadband access networks. NG-OANs are expected to provide services for scenarios, where high requirements are given in terms of long-distance link, high data rate, high stability, and high security.

Table 4.1   Comparison between NG-OANs and 6G RANs.

	NG-OANs	6G RANs
Bandwidth	5 THz	20 GHz
User experience rate	1–10 Gbps	1 Gbps
Peak data rate limit	High	Low
Latency	Downstream latency < 1 ms Upstream latency < 1.5 ms	< 1ms
Transmission distance	> 1000 km	> 100km
Stability	High	Low
Cost	Low	High
Service area	Indoor scenarios	Outdoor mobile scenarios
Security	High	Low
Technology coexistence	Coexistence with 10G PON	Coexistence with 5G
Multiple access technology	TDM/WDM/OFDM/MIMO	MIMO/NOMA
High-data-rate technology	OWC/channel bonding	OWC/THz communication/Quantum communication

Source: Guo et al./IEEE [15].

Conversely, when services are needed in scenarios that are fast-moving or have difficulty in deploying wired networks, 6G RANs are a feasible solution.

4.1.9. 6G Standardization Roadmap

For illustration, Figure 4.3 depicts the 6G standardization roadmap presented in a recent comprehensive 6G survey [16]. The ITU-R Sector has started to develop a new recommendation "IMT Vision for 2030 and beyond" at their March 2021 meeting. This recommendation will define the framework and overall objectives of the future development of IMT for 2030 and beyond, including the role that IMT could play to better serve the needs of the future society, for both developed and developing countries. In addition, ITU-R will hold their world radiocommunication conferences (WRCs) every three to four years to review and, if necessary, revise the radio regulations such as 6G spectrum allocation. Another important 6G standard development organization (SDO) body is the 3rd Generation Partnership

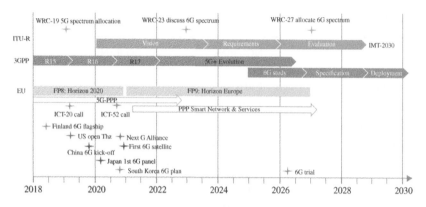

Figure 4.3   6G standardization roadmap including worldwide launched 6G research initiatives. Source: Jiang et al./IEEE [16].

Project (3GPP), which unites seven telecommunications SDOs and produces the reports and specifications that define 3GPP technologies. Currently, 3GPP works on release 18 (Rel-18) to introduce 5G Advanced, which will significantly evolve 5G in the areas of AI and XR, before planning releases on 6G for the year 2027. Other SDO bodies involved in 6G standardization efforts include the following organizations, alliances, and initiatives:

- ETSI
- ITU-T FG-NET-2030
- Next Generation Mobile Networks (NGMN) Alliance
- Alliance for Telecommunications Industry Solutions (ATIS)
- Association of Radio Industries and Businesses (ARIB)
- 3rd Generation Partnership Project (3GPP)
- IEEE Future Networks (FN) Initiative
- Next G Alliance

Note that it is the Next G Alliance where many of the major over-the-top (OTT) Internet players, most notably Apple, Google, Meta (Facebook), Microsoft, and Samsung, are full members. Recall from Section 1.2.2 that these companies, among several others, have already embraced the Metaverse.

4.2. NSF: Next G Research

NSF's view on Next G research and their near-, mid-, and long-term objectives were shown above in Figure 4.1. In this figure, many of the emerging and future key enabling technologies of 6G and Next G networks can be found, ranging from widely studied AI/ML, multi-access edge computing (MEC), joint sensing/communication, ultra-low latency wireless links to zero-trust networking, meta-materials and intelligent surfaces, quantum networks, THz networking, Tactile Internet, and last but not least, the Metaverse.

According to [1], Next G is first and foremost about the convergence of wired and wireless, communications and sensing, HetNets (6G, satellite, WiFi), computation and communications, and software-driven communication systems. In addition, spectrum sharing, resilience, ultra-low latency, wider broadband reach, and intelligence play a central role in Next G. Spectrum sharing is imperative. It can be done through novel means. We have explored only a few options, and there are many more to explore. Spectrum sharing can unleash the power of a shared public resource for greater public good while using it to ensure security, safety, and science. Importantly, according to [1], Next G research is not just 6G, but also advanced WiFi networks, advanced spectrum bands, wireless sensing and imaging. More specifically, Next G research is independent from the various efforts carried out by the different 6G SDOs mentioned in Section 4.1, i.e. Next G includes but is not limited to the specific KPI requirements and topics of interest addressed by 6G SDOs.

4.3. Next G Alliance Roadmap

In February 2022, the Next G Alliance, an ATIS initiative to advance North American wireless technology leadership in 6G and beyond over the next decade, has released their *Roadmap to 6G*.[3] This first release of the report describes the Next G Alliance foundational vision and roadmap in terms of key goals and objectives, the timeline of major milestones on the path to 6G, and key priorities and recommendations that reflect work across the Next G Alliance. The findings and recommendations will be updated in future releases.

4.3.1. Six Audacious Goals

The Next G Alliance has identified six audacious goals that describe the top priorities for North America's contribution and leadership in future global 6G standards, deployments, products, operations, and services. These priorities contemplate both the *societal* and *economic* needs across North America,

---

[3] For free download, please visit: https://roadmap.nextgalliance.org

and the technology strengths that North America will contribute to the rest of the world. Next G Alliance's six audacious goals describe North America's priorities and ambitions for 6G systems:

1. **Trust, Security, and Resilience** should be advanced such that future networks are fully trusted by people, businesses, and governments. Importantly, the Next G Alliance envisions 6G to influence biological, physical, and virtual processes by increasing the *acceleration of digital transformation across society*.

2. An enhanced **Digital World Experience** (DWE) consists of multi-sensory experiences to enable transformative forms of human collaboration. 6G DWEs encompass a variety of multi-sensory experiences that *transform human interactions across physical, digital, and biological worlds*. Innovative human-to-machine interfaces and synergies resulting from machine–machine communications are enablers of more expressive DWE interactions. Inter-personal application DWEs can improve the quality of everyday living (e.g. enabling emotive communications in friends or family interactions), quality of experience (e.g. enhancing shared experiences in multi-user gaming groups), or improve the quality of critical roles (e.g. humanized robotic care). By exploiting MR representations, DWEs aim to allow people to appear anywhere at any time, in time, in any way they choose. DWEs make these opportunities possible by re-shaping today's flat-screen approach with the addition of multi-dimensional, multi-party, and multi-sensory techniques. Social media and enterprise IT businesses in North America are already experimenting with Metaverse strategies, which share many of the cyber-physical, Internet of Senses, and Tactile Internet characteristics associated with 6G DWEs. One national imperative is to initiate policies and programs to scale up and cross-pollinate North America's supply-side ecosystem for 6G DWEs. Broad funding of interdisciplinary research is crucial for DWEs. The fundamental enabling framework for DWEs involves dynamic, multi-sensory, multi-layer representations of the physical world to implement digital twins or mirror worlds. Future applications will rely on a merge of digital and physical worlds to create a wide variety of highly immersive experiences through deeper levels of human–computer interaction. The combination of these technologies is expected to yield *human and machine experiences unthinkable with previous generations*.

3. **Cost-Efficient Solutions** should span all aspects of the network architecture, including devices, wireless access, cell-site backhaul, overall distribution, and energy consumption.

4. **Distributed Cloud and Communications Systems** built on virtualization technologies will increase flexibility, performance, and resiliency for key use cases such as MR, ultra-reliable and low-latency communication (URLLC) applications, interactive gaming, and multi-sensory applications.

5. An **AI-Native Network** is needed to increase the robustness, performance, and efficiencies of wireless and cloud technologies against more diverse traffic types, ultra-dense deployment topologies, and more challenging spectrum situations.

6. **Sustainability** related to energy efficiency and the environment must be at the forefront of decisions throughout the life cycle, toward a goal of achieving IMT carbon neutral by 2040. Advances will fundamentally change how electricity is used to support next-generation communications and computer networks, while strengthening the role that information technology plays in protecting the environment. Scientific reports have raised global awareness around climate change in the past five years and convinced the United States, Canada, and most other worldwide governments to commit to keeping global temperature increases to well below 2°C above pre-industrial levels. The roadmap effort will examine ways in which 6G can contribute to our "green" future, propose technological initiatives to gain that future, and help us realize the green, energy-efficient information technology ecosystem that will benefit the world. Climate challenges are expected to seriously disrupt business as usual and change the way citizens worldwide live their lives. As more companies commit to adopting corporate and social responsibility (CSR) strategies that address environmental and social issues, many companies have a sustainability program in place that commits to reducing their energy consumption and environmental impact.

4.3.2. Societal and Economic Needs

According to the Next G Alliance roadmap, there is a unique opportunity to consider how to address critical North American societal and economic challenges, as we begin defining 6G and identify the key infrastructure capabilities. More specifically, digital equity, trust, sustainability, economic growth, and quality of life are key pillars in addressing the *interdependencies between human and technological evolution*. Digital equity should be understood as a requirement to achieve affordability, accessibility, and geographic availability for each user. Key components of trust include security, data privacy, and resiliency. The social and economic aspects of sustainability are closely connected to a broader set of environmental concerns across the entire value chain of

6G telecommunication infrastructure and technologies, encompassing raw materials, supply chain, operation, and disposal. Furthermore, 6G-enabled services will be important to improving the quality of life in North America and its local communities. Note that the five groups of common outcomes that the Next G Alliance recommends prioritizing for North America nicely align with the United Nations' Sustainable Development Goals (SDGs).

Importantly, there is a *symbiotic relationship* between technology and a population's societal and economic needs. As *technology shapes human behavior* and lifestyles, those *needs shape technological evolution*. Issues around digital equity, trust, sustainability, economic growth, and quality of life are both societal and economic imperatives. According to the Next G Alliance roadmap, these are therefore outcomes for 6G systems to target. Achieving these outcomes requires an integration of these social and economic issues throughout the full 6G lifecycle of research and development.

4.3.3. The Four Fundamental Areas for 6G Applications and Use Cases

The Next G Alliance is exploring new opportunities that are anticipated to arise from 6G applications. The use cases behind these applications and markets can be summarized into four foundational areas, as illustrated in Figure 4.4. Note that the mapping of use cases to the fundamental areas is not exclusive since some areas have a certain degree of commonality. The four foundational areas for 6G applications and use cases are defined as follows:

1. **Living**: With the enhanced capabilities, sensing, and actuation deeply integrated in the 6G fabrics, *humans are expected to be the ultimate*

Figure 4.4   Next G Alliance Roadmap to 6G: the four foundational areas for 6G applications and use cases. Source: Aunging/Adobe Stock.

beneficiaries of 6G. For example, ambient intelligence delivering seam-less immersive experiences regardless of the point of consumption will open new doors for development of human-centric technologies. This will not only improve our daily lives, but also *positively influence human behaviors, in turn advancing the societies* they create.

2. **Experience**: One of the Next G Alliance's audacious goals is 6G DWEs. These use cases target services in the field of human–machine interactions, technology-mediated human-to-human interactions, physical and psychological health assistance, MR entertainment, and real-time interactive gaming with physical interactions, powered with MR content and XR-enriched capabilities. Although some experience use cases can be powered by 5G networks, there is a major expectation to refine or improve the feasibility and 6G user experience through a more advanced network building on holographic projections, which will be crucial for user experience.

3. **Critical**: The Next G Alliance has assembled a library of use cases that illustrate ways to advance the quality of technology in and around health, health care, manufacturing, agriculture, and public safety with the use of robotics. Nearly every use case identifies a role for robotic–human interaction and the use of robots as a *complement*, rather than a wholesale replacement for human involvement. Digital twin and similar capabilities will *augment the capabilities*, allowing humans to interact naturally with robots in the digital space and then to *initiate actions in the physical world*.

4. **Societal Goals**: With the advent of 6G, there is a unique opportunity to address the goals of *digital equity* and *social sustainability* through applications leveraging virtual reality (VR)/augmented reality (AR) designed to enable and encourage self-sufficiency, participation, and collaboration. High-level functional requirements in support of sustainable, equitable, and inclusive societies include the elimination of digital divide technical barriers and the need for basic-level tac-tile/visual VR/XR type of services at a lower price point to support physically challenged subscribers' wearables or assistance devices (e.g. voice, tactile, visual).

It is important to note that the recommended actions of the roadmap focus on how they would influence humans living in the society. Spanning from the quality of everyday living and the quality of truly immersive experience through to the quality of critical role-playing, promising 6G use cases and applications are expected to provide confidence on how new technology solutions can eventually influence the quality of humans' living in line with the Next G Alliance's audacious goals, including societal priorities for North America. Toward this end, 6G applications will improve the quality of everyday living, enable new experiences, support mission critical needs, and become a key factor in attaining societal goals.

References

1. A. Sprintson. An Overview of NSF's Support for Next G Research. In *Proc., First IEEE Next G Summit*, June 2022.

2. Ericsson. 6G — Connecting a cyber-physical world: A research outlook toward 2030. pages 1–31. White Paper GFTL-20:001402, February 2022.

3. S. Yrjölä, P. Ahokangas, and M. Matinmikko-Blue. Value creation and capture from technology innovation in the 6G era. *IEEE Access*, 10:16299–16319, February 2022.

4. D. C. Nguyen, M. Ding, P. N. Pathirana, A. Seneviratne, J. Li, D. Niyato, O. Dobre, and H. V. Poor. 6G Internet of Things: a comprehensive survey. *IEEE Internet of Things Journal*, 9(1):359–383, January 2022.

5. F. Liu, Y. Cui, C. Masouros, J. Xu, T. X. Han, Y. C. Eldar, and S. Buzzi. Integrated sensing and communications: toward dual-functional wireless networks for 6G and beyond. *IEEE Journal on Selected Areas in Communications*, 40(6):1728–1767, June 2022.

6. C. Wang and A. Rahman. Quantum-enabled 6G wireless networks: opportunities and challenges. *IEEE Wireless Communications*, 29(1):58–69, February 2022.

7. A. H. Khan, N. U. Hassan, C. Yuen, J. Zhao, D. Niyato, Y. Zhang, and H. V. Poor. Blockchain and 6G: the future of secure and ubiquitous communication. *IEEE Wireless Communications*, 29(1):194–201, February 2022.

8. A. Kalla, C. D. Alwis, S. P. Gochhayat, G. Gür, M. Liyanage, and P. Porambage. Emerging directions for blockchainized 6G. *IEEE Consumer Electronics Magazine*, pages 1–7, April 2022. IEEE Xplore Early Access.

9. W. Tong and G. Y. Li. Nine challenges in artificial intelligence and wireless communications for 6G. *IEEE Wireless Communications*, pages 1–10, May 2022. IEEE Xplore Early Access.

10. F. Zhou, W. Li, Y. Yang, L. Feng, P. Yu, M. Zhao, X. Yan, and J. Wu. Intelligence-endogeneous networks: innovative network paradigm for 6G. *IEEE Wireless Communications*, 29(1):40–47, February 2022.

11. K. B. Letaief, Y. Shi, J. Lu, and J. Lu. Edge artificial intelligence for 6g: vision, enabling technologies, and applications (invited paper). *IEEE Journal on Selected Areas in Communications*, 40(1):5–36, January 2022.

12. I. F. Akyildiz, C. Han, Z. Hu, S. Nie, and J. M. Jornet. Terahertz band communication: an old problem revisited and research directions for the next decade (invited paper). *IEEE Transactions on Communications*, 70(6):4250–4285, June 2022.

13. H.-J. Song and N. Lee. Terahertz communications: challenges in the next decade (invited paper). *IEEE Transactions on Terahertz Science and Technology*, 12(2):105–117, March 2022.

14. R. Sabella, D. V. Plant, H. Chen, A. Bogoni, and V. Stojanovic. Guest editorial: optical systems and technologies for 6G mobile networks. *IEEE/OPTICA Journal of Lightwave Technology*, 40(2):336–338, January 2022.

15. H. Guo, Y. Wang, J. Liu, and N. Kato. Super-broadband optical access networks (OANs) in 6G: vision, architecture, and key technologies. *IEEE Wireless Communications*, pages 1–11, May 2022. IEEE Xplore Early Access.

16. W. Jiang, B. Han, M. A. Habibi, and H. D. Schotten. The road towards 6G: a comprehensive survey. *IEEE Open Journal of the Communications Society*, 2:334–366, February 2021.

CHAPTER 5

6G Post-Smartphone Era: XR and Hybrid-Augmented Intelligence

"The Internet will disappear."

ERIC SCHMIDT
Former Executive Chairman of Google
(1955–present)

5.1. XR in the 6G Post-Smartphone Era

At the 2015 World Economic Forum, Eric Schmidt famously stated that "the Internet will disappear" given that there will be so many things that we are wearing and interacting with that we won't even sense the Internet, though it will be part of our presence all the time. Although this first might sound a bit surprising, it is actually what profound technologies do in general. In "The Computer for the 21st Century," Mark Weiser argued that the most profound technologies are those that disappear. They weave themselves into the fabric of everyday life until they are indistinguishable from it [1].

An interesting recent approach to make the Internet disappear is the so-called *naked world* vision that aims at paving the way to the Internet of no things by offering all kinds of human-intended services without owning or carrying any type of computing or storage devices [2]. The term "Internet of no things" was coined by Demos Helsinki founder of Roope Mokka in 2015. The term nicely resonates with Eric Schmidt's aforementioned statement.

6G and Onward to Next G: The Road to the Multiverse, First Edition. Martin Maier.
© 2023 The Institute of Electrical and Electronics Engineers, Inc.
Published 2023 by John Wiley & Sons, Inc.

The naked world envisions Internet services to appear from the surrounding environment when needed and disappear when not needed. The transition from the current gadgets-based Internet to the Internet of no things is divided into three phases that starts from bearables (e.g. smartphone), moves toward wearables (e.g. Google and Levi's smart jacket or Amazon's announced voice-controlled echo loop ring, glasses, and earbuds), and then finally progresses to the last phase of so-called nearables. Nearables denote nearby surroundings or environments with embedded computing/storage technologies and service provisioning mechanisms that are intelligent enough to learn and react according to user context and history in order to provide user-intended services. According to [2], their successful deployment is challenging not only from a technological point of view but also from a business and social mindset perspective due to the required user acceptability and trust.

Some of the most interesting 5G applications – most notably, virtual reality (VR) and the so-called Tactile Internet based on haptic communications – seem to evolve in the same direction. To see this, note that according to [3], the VR systems will undergo three evolutionary stages, similar to the aforementioned Internet of no things. The first evolutionary stage includes current VR systems that require a wired connection to a PC or portable device because current wireless systems cannot cope with the massive amount of bandwidth and latency requirements of VR. The PC or portable device in turn is connected to the central cloud and the Internet via backhaul links. At the second evolutionary stage, the VR devices are wirelessly connected to a fog/edge server located at the base station (BS) for local computation and caching. The third and final evolutionary stage envisions ideal (fully interconnected) VR systems, where no distinction between real and virtual worlds are made in human perception. In addition, according to [3], the growing number of drones, robots, and self-driving vehicles will take cameras to places humans could never imagine reaching. Similarly, the Tactile Internet, specified within the IEEE P1918.1 standards working group, allows for the tactile steering and control of not only virtual but also real objects (e.g. teleoperated robots) as well as processes. Thus, the Tactile Internet may be viewed as an extension of immersive VR from a virtual to a physical environment.

The above discussion shows that future fully interconnected VR systems and the Tactile Internet seem to evolve toward common design goals. Most notably, the boundary between virtual (i.e. online) and physical (i.e. offline) worlds is to become increasingly imperceptible, whereas both digital and physical capabilities of humans are to be extended via edge computing variants, ideally with embedded AI capabilities. According to the inaugural report "Artificial Intelligence and Life in 2030" of Stanford University's recently launched *One Hundred Year Study on AI (AI100)*, an increasing focus on developing systems that are human-aware is expected over the next 10–15 years.

In this chapter, we elaborate on how the Internet of no things with its underlying human-intended services may serve as a useful stepping stone toward realizing the far-reaching vision of future 6G networks, ushering in the 6G post-smartphone era. After briefly putting the Internet of no things in 6G perspective, we explain the reality–virtuality continuum and exploit the Multiverse for the design of advanced extended reality (XR) experiences, ranging from conventional VR to more sophisticated cross-reality environments known as third spaces. We then elaborate on the recently emerging *invisible-to-visible* (I2V) technology concept, which we use together with other key enabling network technologies to tie both online and offline worlds closer together in an Internet of no things and make it "see the invisible" through the awareness of nonlocal events in space and time.

5.1.1. Internet of No Things: The Gadget-Free Internet

Letaief et al. [4] provided a roadmap to 6G, which envisions that, in contrast to previous generations, 6G will be transformative and will revolutionize the wireless evolution from "connected things" to "connected intelligence." According to [5], 6G will play a significant role in advancing Nikola Tesla's prophecy that "when wireless is perfectly applied, the whole Earth will be converted into a huge brain." Toward this end, Strinati et al. [5] argue that 6G will provide an information and communications technologies (ICT) infrastructure that enables end-users to perceive themselves as surrounded by a huge artificial brain offering virtually zero-latency services, unlimited storage, and immense cognitive capabilities. In 6G, there is also a strong notion that the nature of mobile terminals will change, whereby smart cars and intelligent mobile robots are anticipated to play a more important role [6].

6G is anticipated to allow for the inclusion of additional human sensory information. International Telecommunication Union-Telecommunication Sector Focus Group Technologies for Network (ITU-T FG NET)-2030 was established in July 2018 to study and advance the capabilities of the networks for the year 2030 and beyond. Among others, FG NET-2030 envisions user experiences to go from well-explored, audio-visual communications to the delivery of all five human senses as well as *other senses* in line with the IEEE Digital Senses Initiative. David and Berndt [7] advocate that 6G should embrace a new mode of thinking from the get-go by including social awareness and understanding the social impact of advanced technologies. They argue that deepened personalization of 6G services that could *predict future events* for the user and provide good advice would certainly be appreciated.

Finally, Saad et al. [8] observed that the ongoing deployment of 5G cellular systems is exposing their inherent limitations compared to the original premise of 5G as an enabler for the Internet of everything (IoE). They argue that 6G should not only explore more spectrum at high-frequency bands but, more importantly, converge driving technological trends, thereby ushering in the 6G post-smartphone era. Their bold, forward-looking research agenda intends to serve as a basis for stimulating more out-of-the-box research that will drive the 6G revolution. Specifically, they claim that there will be the following four driving applications behind 6G: (i) multi-sensory XR applications, (ii) connected robotics and autonomous systems, (iii) wireless brain–computer interaction (a subclass of human–machine interaction (HMI)), and (iv) blockchain and distributed ledger technologies (DLT). Among other 6G driving trends and enabling technologies, they emphasize the importance of haptic and empathic communications, edge artificial intelligence (AI), the emergence of smart surfaces/environments and new human-centric service classes, as well as the end of the smartphone era, given that smart wearables are increasingly replacing the functionalities of smartphones. They also expect that research on the *quantum realm* will intersect with 6G toward its end of standardization.

The Internet of no things with its underlying human-intended services and nonlocal extension of human "sixth-sense" experiences in both space and time may serve as a useful stepping stone toward realizing the far-reaching 6G vision, as explained in technically greater detail in the remainder of the chapter.

5.1.2. Extended Reality (XR): Unleashing Its Full Potential

In this section, we further elaborate on the recently emerging term XR and how its full potential can be unleashed via the Multiverse.

According to Qualcomm, XR will be the next-generation mobile computing platform that brings the different forms of reality together in order to realize the entire reality–virtuality continuum of Figure 5.1 for the extension

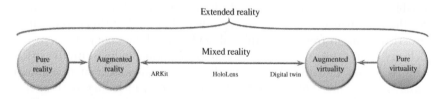

Figure 5.1   The reality–virtuality continuum, ranging from pure reality (offline) to pure virtuality (online).

of human experiences, including the support of HMI. In fact, according to a recent application binary interface (ABI) Research and Qualcomm study, some of the most exciting XR use cases include remotely controlled devices and the Tactile Internet [9].

The reality–virtuality continuum ranges from pure reality (offline) to pure virtuality (online), as created by VR. Both reality and virtuality may be augmented, leading to augmented reality (AR) on one side of the continuum and augmented virtuality (AV) on the other. AR enables the live view of a physical, real-world environment, whose elements are augmented by computer-generated perceptual information, ideally across multiple sensory modalities. In doing so, AR alters one's perception of the real-world environment, as opposed to VR, which replaces the real-world environment with a simulated one. Conversely, AV occurs in a virtual environment, where a real object is inserted into a computer-generated environment.

A good AR example is Shopify's ARKit, which allows smartphones to place 3D models of physical items and see how they would look in real life. To do so, AR overlays virtual objects/images/information on top of a real-world environment. An illustrative AV example is an aircraft maintenance engineer, who is able to visualize a real-time model, referred to as digital twin, of an engine that may be thousands of kilometers away. The term *mixed reality* (MR) includes AR, AV, and mixed configurations thereof, blending representations of virtual and real-world elements together in a single-user interface. MR helps *bridge the gap between real and virtual environments*, whereby the difference between AR and AV reduces to where the user interaction takes place. If the interaction happens in the real world, it is considered AR. By contrast, if the interaction occurs in a virtual space, it is considered AV. The flagship MR device is Microsoft's HoloLens. XR is the umbrella term that spans the entire reality–virtuality continuum and its different types of physical and/or VR, including AR, MR, and AV. The areas where most industries apply XR is in remote guidance systems for performing complex tasks such as maintenance and assembly [10].

5.1.3. Invisible-to-Visible (I2V) Technologies

Recall from above that future fully interconnected VR systems will leverage on the growing number of drones, robots, and self-driving vehicles. A very interesting example of future connected-car technologies that merges real and virtual worlds to help drivers "see the invisible" is Nissan's recently unveiled invisible-to-visible (*I2V*) technology concept [11]. I2V creates a three-dimensional immersion connected-car experience that is tailored to the driver's interests by changing how cars are driven and integrated into society. More specifically, by merging information from sensors outside and inside the vehicle with data from the cloud, I2V enables the driver and passengers not only to track the vehicle's immediate surroundings but also to anticipate what is ahead, e.g. what is behind a building or around the corner. Although the initial I2V proof-of-concept demonstrator used AR headsets (i.e. wearables), Nissan envisions to turn the windshield of future self-driving cars into a portal to the virtual world, thus finally evolving from wearables to nearables, as discussed above in the context of the Internet of no things.

I2V is powered by Nissan omnisensing technology, a platform originally developed by the video gaming company Unity Technologies, which acts as a hub gathering real-time data from the traffic environment and from the vehicle's surroundings and interior to anticipate when people inside the vehicle may need assistance. The technology maps a 360 deg virtual space and gives guidance in an interactive, human-like way, such as through avatars that appear inside the car. It can also connect passengers to people in the *Metaverse* virtual world that is shared with other users. In doing so, people may appear inside the car as AR avatars to provide assistance or company. For instance, when visiting a new place, I2V can search within the Metaverse for a knowledgeable local guide. The information provided by the guide may be stored in the cloud such that others visiting the same area can access it or may be used by the onboard AI system for a more efficient drive through local areas. Alternatively, the driver may book a professional driver from the Metaverse, who appears as a virtual chase car in the driver's field of view to show the best way and improve driving skills, just like in a video game.

Clearly, I2V opens up endless opportunities by tapping into the virtual world. In fact, the IEEE P1918.1 standards working group, briefly mentioned above, highlights several key use cases of the Tactile Internet, including not only the automative control of connected/autonomous driving via virtual avatars but also the remote control of physical robots. According to [12], the vastly progressing smart wearables such as exoskeletons and VR/AR devices effectively create real-world avatars, i.e. tactile robots connected with human operators (HOs) via smart wearables, as a central physical embodiment of the Tactile Internet. More specifically, Haddadin et al. [12] argue that the Tactile Internet creates the new paradigm of an immersive coexistence between humans and robots in order to achieve tight physical human–robot interaction (pHRI) and entanglement between man and machine in future locally connected human–avatar/robot collectives. Assistive exoskeletons are thereby envisaged to become an important element of the Tactile Internet in that they extend user capabilities and supplement or replace some form of function loss, e.g. lifting heavy objects or rehabilitation systems for people with spinal cord injury. In addition, many studies have shown that the physical presence of robots benefited a variety of social interaction elements such as persuasion, likeability, and trustworthiness. Thus, leveraging these beneficial characteristics of social robots represents a promising solution toward addressing the user acceptability and trust issues of nearables mentioned above.

In the following, we build on the I2V technology concept and explore how emerging multisensory XR technologies in conjunction with AI-enhanced multi-access edge computing (MEC), intelligent mobile robots, and blockchain technologies may be combined to usher in the Internet of no things as an important stepping stone toward realizing the 6G vision outlined in Section 5.1.1.

5.1.4. Extrasensory Perception (ESP)

Let our point of departure be Joseph A. Paradiso's pioneering work on *extrasensory perception* (ESP) in an Internet of things (IoT) context at MIT Media Lab [13]. In a sensor-driven world, network-connected sensors embedded in anything function as extensions of the human nervous system and enable us to enter the long-predicted era of ubiquitous computing, as envisioned by Mark Weiser more than a quarter of century ago. Dublon and Paradiso [13] showed that network-connected sensors and computers make it possible to virtually travel to distant environments and "be" there in real time. Interestingly, the authors concluded that future technologies will fold into our surroundings that help us get our noses off the smartphone screens and back into our environments, thus making us more (rather than less) present in the world around us. Clearly, this human-centric outlook on future technologies may materialize in the 6G post-smartphone era.

Recall from Section 5.1.2 that XR will be the next-generation mobile computing platform for the extension of human experiences, including the support of HMI. In Chapter 1, we briefly introduced our proposed Internet of No Things architecture, which we now use for demonstrating two illustrative examples of ESP. As shown in Figure 1.1, the underlying physical network infrastructure of the architecture consists of a fiber backhaul shared by wireless local area network (WLAN) mesh portal points (MPPs) and cellular BSs that are collocated with optical network units (ONUs), which in turn are connected to the central optical line terminal (OLT) of the fiber backhaul. In addition to conventional audio-visual and data communications, users may engage in nonlocal teleoperation between a HO and teleoperator robot (TOR), which are both physical, i.e. offline, entities, as depicted in Figure 1.1. It is worthwhile to mention that AI-enhanced MEC may be exploited to decouple haptic feedback between pairs of TOR and HO from the impact of extensive propagation delays by forecasting delayed or lost haptic feedback samples. This enables humans to perceive remote task environments in real time at a 1 ms granularity. For further information, we refer the interested reader to [14].

Note that an interesting phenomenon for changing behavior in an online virtual environment is the so-called "Proteus effect," where the behavior of individuals is shaped by the characteristics and traits of their virtual avatars, especially through interaction during inter-avatar events (see also Figure 1.1). Toward this end, we will exploit AI-enhanced MEC to realize persuasive computing, as described in more detail shortly.

5.1.5. Mimicking the Quantum Realm: Nonlocal Awareness of Space and Time

As an illustrative example of advanced XR experiences, we study the delivery of extrasensory human perceptions, i.e. senses other than the five human senses, as envisioned by the IEEE Digital Senses Initiative and ITU-T FG-NET-2030 (see Section 5.1.1).

It is interesting to note that the term "ESP" actually refers to a widely known phenomenon that allows humans to have nonlocal experiences in space and time. According to Wikipedia, ESP is also called *sixth sense*, which includes claimed reception of information not gained through the recognized five physical senses, but sensed with the mind. There exist different types of ESP, including clairvoyance (i.e. viewing things or events at remote locations) and precognition (i.e. viewing future events before they happen). While clairvoyance may be viewed as the ability to perceive the hidden present, precognition is a forecast (not prophecy) of events to come about in the future unless one does something to change them based on the perceived information. In contemporary physics, there exists the so-called "principle of nonlocality," also referred to as *quantum-interconnectedness of all things* by quantum physicists such as David Bohm, which transcends spatial and temporal barriers [15]. Nonlocality occurs due to the phenomenon of entanglement, where a pair of particles have complementary properties when measured, and might be the cause of ESP.

Phenomena such as entanglement also play an important role in the nascent Quantum Internet [16]. The Quantum Internet consists of both classical and quantum links interconnecting remote quantum devices. With respect to quantum communication resources, it seems attractive to utilize existing optical fiber networks. However, it is still an open problem to determine whether it is feasible to utilize a single link, e.g. a single optical fiber, for both quantum and classical communications, such that existing network infrastructures can be exploited without the need for additional infrastructures. Hence, from a communication engineering perspective, the design of the Quantum Internet is not an easy task at all since it is governed by the laws of quantum phenomena with no counterpart in classical networks, which impose serious constraints on the network design. A key strategy for transmitting information in the Quantum Internet is teleportation. Quantum teleportation provides an invaluable strategy for transmitting so-called quantum bits (qubits) without either the physical transfer of the particle storing the qubit or the violation of the quantum mechanics principles.

According to [16], the Quantum Internet is probably still a concept far from real-world implementation. In addition, the quantum teleportation process, which represents the core communication functionality of the Quantum Internet, is gravely affected by a number of quantum imperfections that arise during the quantum teleportation process from a communication engineering perspective [17]. Despite the fact that the Quantum Internet might pave the way for the Internet of the future, there is a substantial amount of frontier-research required for tackling the challenges and open problems associated with it. By contrast, with the advent of advanced XR technologies it might be easier to *mimic the Quantum Internet* instead of actually building it, as explained in more detail next.[1]

Note that despite reports based on anecdotal evidence, there has been no convincing scientific evidence that ESP exists after more than a century of research. However, instead of rejecting ESP as pseudoscience, in this chapter, we argue that with the emergence of XR it might become possible to disrupt the old impossible/possible boundary and mimic the quantum realm. Toward this end, we are going to design the following two advanced XR experiences that transverse the boundary between the Multiverse realms in order to realize awareness of nonlocal events in space and time.

5.1.5.1. Precognition

To achieve precognition, we extend our AI-enhanced MEC-based haptic feedback sample forecasting scheme in [14] for realizing persuasive computing. Recall from above that in nonlocal teleoperation the HO and TOR are physical entities, i.e. both reside in the Multiverse realm reality characterized by Space, Time, and Matter. In addition, we let the HO have access to the Multiverse realm AV, characterized by No-Space, Time, and No-Matter, by observing a digital twin of the remote TOR via a wearable HMD.

Our AI-enhanced MEC forecasting scheme was trained by using haptic traces obtained from application-specific teleoperation experiments. We showed in [14] that a high forecasting accuracy (mean squared error below 1‰) can be achieved in the considered scenarios. In general, however, the training may become irrelevant in changing or unstructured real-world environments, resulting in a decreased forecasting accuracy. How can the

---

[1] Still unsettled is the discussion about whether the brain is a natural quantum computer or not. Nevertheless, it is worthwhile to mention that there are many theories that in some way relate the brain to quantum physics, where quantum effects play some kind of role in the brain.

HO know when or even before this happens and be persuaded to make an informed decision?

To quantify the decreasing effectiveness, our AI-enhanced MEC computes the metric regret, which measures the future regret the HO will have after *blindly* relying on a presumably intact haptic feeback sample forecasting scheme. We define regret as the difference between the achievable and the optimum physical task execution times of the TOR. Note that the metric regret is used to influence the HO's decision to abort the teleoperation before unintended consequences might occur. It is displayed in his head-mounted wearable to "make him see" the AI becoming less trustworthy. We will highlight some illustrative results below.

5.1.5.2. Eternalism

Next, we consider also the transition from the Time to No-Time dimension of the Multiverse. In physics, the two most important theories on the nature of time have been *presentism* and *eternalism*. Presentism states that only the present is real. By contrast, eternalism states that the past and future are as equally real as the present. Under eternalism, "now" is to time as "where" is to space, whereby time is a dimension much like space, one in which the past and future are as real as locations north and south (i.e. unlike presentism, eternalism thus lends itself to time travel). Today, most physicists view eternalism as the order of time [18].

Figure 5.2 illustrates our experimental set-up for demonstrating eternalism in locally connected human–avatar/robot collectives. The human engages in embodied communication with Pepper, SoftBank Mobile's most advanced humanoid robot, via voice, gesture, and Pepper's built-in Android tablet. We use an Oculus Rift VR headset to let the human also access the virtual avatar of the robot. A user profile is maintained to record each human–robot interaction. In addition, we exploit IBM Watson's empathic AI services, most notably, IBM Watson's tone analyzer for detecting emotions in written text exchanged during human–robot/avatar online communication. As blockchain of choice, we deploy Ethereum to interconnect all HOs, real and virtual robots, and empathic AI services in a DAO, one of Ethereum's salient features, to share skills and help solve complex problems faced during human–robot–avatar interactions. (DAO and Ethereum will be explained in technically greater detail below in Section 5.2.) MEC-based cloud computing is used for offloading compute-intensive blockchain transactions, e.g. mining, from resource-limited robots. In the subsequent section, we highlight a use case of exploiting VR and empathic AI to *make emotions visible* and nudge the human toward experiencing eternalism.

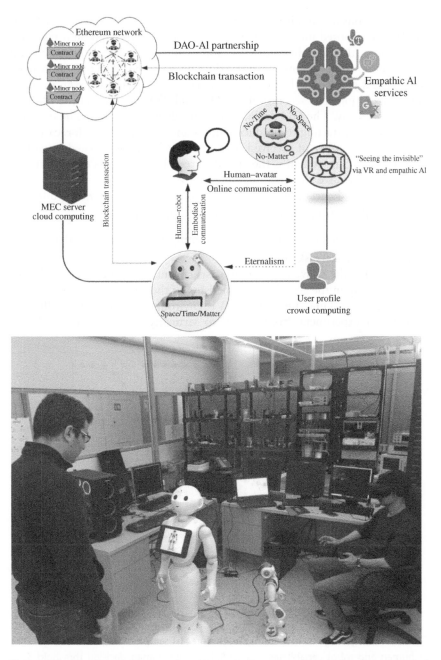

Figure 5.2   Experimental set-up for demonstrating eternalism in locally connected human–avatar/robot collectives. Source: Maier et al. (2020) © 2020 IEEE.

Let us consider an HO–TOR pair carrying out a given physical task that can be decomposed into 100 operations. To achieve the optimum task execution time, the AI-enhanced MEC forecasting scheme outsources certain operations to another crowdsourced HO, who is located 20 seconds away from the physical task point. Let f_H and f_R denote the capability (given in number of operations per second) of the HO and TOR to execute the physical task, respectively. The HO decides to abort teleoperation, when he observes the digital twin starting to produce failures and the ratio of misforecast samples to total number of received haptic feedback samples exceeds a certain threshold S_H. Subsequently, the crowdsourced HO traverses to the physical task point to finalize all remaining operations.

Figure 5.3 depicts the regret vs. misforecast sample rate λ_f for different ratio $\frac{f_H}{f_R}$ and S_H. It highlights the beneficial role of crowdsourcing a capable assistant HO with increased $\frac{f_H}{f_R}$ in compensating for an unreliable AI and completing the physical task failure-free.

Note that the above digital twin is synchronized with the remote TOR, both operating in the actual time dimension of the multiverse. Next, we also tap into the No-Time dimension of VR environments during the following time travel experiment from reality to virtuality. The experiment lasted

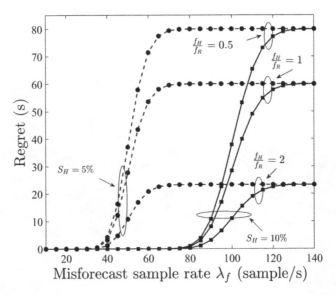

Figure 5.3   Regret (given in seconds) vs. misforecast sample rate λ_f for different ratios of human and robot capabilities $\frac{f_H}{f_R} \in \{0.5, 1, 2\}$ with human decision threshold $S_H \in \{5\%, 10\%\}$. Source: Maier et al. (2020) © 2020 IEEE.

15 minutes and was repeated five times, each time involving a different student. In the initial reality part, the student first engages with Pepper for an interactive audio–visual tour of INRS (the students' university). Subsequently, the student is given the opportunity to ask Pepper any arbitrary question about INRS, whereby Pepper's responses are provided by a remote HO via speech-to-text and text-to-speech conversion. Next, Pepper invites the student to continue the experiment in the virtuality part, where the student can virtually walk through INRS guided by an avatar acting as an omniscient oracle. The oracle relies on a remote HO, who is able to monitor the student's detected emotions in real time. By leveraging on the "Proteus effect" experienced in inter-avatar events, the oracle gives advice to the student on how to gradually reach a desirable future situation at INRS, which is characterized by higher levels of confidence and emotional engagement.

Figure 5.4 shows the average empathic AI score of the four positive emotions detected by IBM Watson's tone analyzer during the various

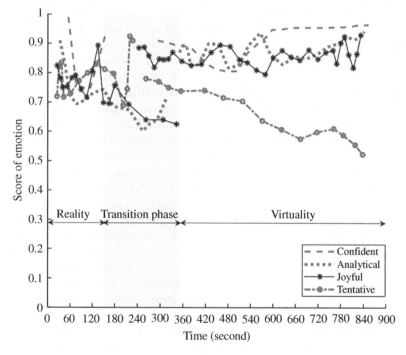

Figure 5.4   Average empathic AI score of four different positive emotions experimentally detected during time travel from reality to virtuality. Source: Maier et al. (2020) © 2020 IEEE.

human–robot/avatar speech-to-text-to-speech exchanges of the experiment. It clearly illustrates that the students become increasingly more confident and emotionally less tentative after transiting from reality to virtuality.

5.2. Hybrid-Augmented Intelligence

Today's Internet is ushering in a new era. While the first generation of digital revolution brought us the Internet of information, the second generation – powered by decentralized blockchain technology – is bringing us the *Internet of value*, a true peer-to-peer platform that has the potential to go far beyond digital currencies and record virtually everything of value to humankind in a distributed fashion without powerful intermediaries. Arguably more importantly, though, according to Don and Alex Tapscott the blockchain technology enables trusted collaboration that can start to change the way wealth is distributed as people can share more fully in the wealth they create. As a result, decentralized blockchain technology helps create platforms for distributed capitalism and a more inclusive economy [19].

The fundamental concepts and potential of blockchain technologies for society and industry in general have been described comprehensively in various existent tutorials, e.g. [20]. In particular, there has been a growing interest in adapting blockchain to the specific needs of the IoT in order to develop a variety of blockchain-based Internet of things (BIoT) applications, ranging from smart cities and Industry 4.0 to financial transactions and farming, among others [21]. Toward this end, Fernández-Caramés and Fraga-Lamas [22] pointed out the important role of smart contracts, which are defined as pieces of self-sufficient decentralized code that are executed autonomously when certain conditions are met, whereby *Ethereum* was one of the first blockchains using smart contracts. Ethereum can be described as a blockchain with built-in programming language (Solidity) and a consensus-based virtual machine running globally as Ethereum virtual machine (EVM). The use of Ethereum allows users to write and run their own code on top of the network. By updating the code, users are able to modify the behavior of IoT devices for simplified maintenance and error correction. Beside well-known BIoT problems such as hosting a blockchain on resource-constrained IoT devices, low transaction rates, and long block creation times, Fernández-Caramés and Fraga-Lamas [22] identified several significant challenges beyond early BIoT developments and deployments that will need further investigation. Apart from technological challenges, e.g. access control and security, the authors concluded that shaping the regulatory environment, e.g. *decentralized ownership*, is one of the biggest issues to unlock the potential of BIoT for its broader use.

Recently, initial studies have begun to address some of the aforementioned shortcomings of BIoT. In [23], resource-constrained IoT devices were released from computation-intensive tasks by offloading the mining work (i.e. creating/appending/monitoring blockchain transactions) onto more powerful edge computing resources such as cloudlets. The proposed *EdgeChain* was built on the Ethereum platform and uses smart contracts to monitor and regulate the behavior of IoT devices based on their resource usage and activities. Since all activities are recorded in the blockchain, it is difficult for malicious nodes to cause sustained damage or run away with no traces. Furthermore, to tackle the critical access control issue of preventing BIoT resource access from unauthorized entities, Zhang et al. [24] exploited the Ethereum smart contract platform to achieve various access control methods. Specifically, gateways were used to act as BIoT service agents for their respective cluster of local resource-constrained IoT devices by storing their blockchain accounts and using them to execute smart contracts on their behalf. The proposed smart contract-based framework consists of multiple access control contracts (ACCs). Each ACC maintains a misbehavior list for each BIoT resource, including details and time of the misbehavior as well as the penalty on its subject, e.g. blocking access requests for a certain period of time. Further, in addition to a register contract, the framework involves the so-called *judge contract* (JC), which implements a certain misbehavior judging method. After receiving the misbehavior reports from the ACCs, the JC determines the penalty on the corresponding subjects and returns the decisions to the ACCs for execution.

Many additional BIoT studies considered Ethereum as the blockchain of choice. For instance, the architectural issues for realizing BIoT services were investigated in greater detail in [25]. In a preliminary study using a smart thing renting service as an example of BIoT service, the authors compared the following four different architectural styles based on Ethereum: (i) fully centralized (cloud without blockchain), (ii) pseudo distributed things (physically located in central cloud), (iii) distributed things (directly controlled by smart contract), and (iv) fully distributed. The preliminary results indicate that a fully distributed architecture, where a blockchain endpoint is deployed on the end-user device, is superior in terms of robustness and security. Further, the various perspectives for integrating secure elements in Ethereum transactions were discussed in [26]. A novel architecture for establishing trust in Ethereum transactions exchanged by smart things was presented, which enables remote and safe digital signatures by using the well-known elliptic curve digital signature algorithm (ECDSA).

Despite the recent progress, the salient features that set Ethereum aside from other blockchains remain to be explored in more depth, including their symbiosis with other emerging key technologies such as AI and robots apart from decentralized edge computing solutions. A question of particular interest hereby is how decentralized blockchain mechanisms may be leveraged to let emerge new hybrid forms of collaboration among individuals, which haven't been entertained in the traditional market-oriented economy dominated by firms rather than individuals [20]. Of particular interest will be Ethereum's concept of the DAO. In fact, in their latest book on how to harness our digital future [27], Andrew McAfee and Erik Brynjolfsson speak of "The Way of The DAO" that may substitute a technology-enabled crowd for traditional organizations such as companies.

In the following, we focus on the so-called Tactile Internet, which is supposed to be next leap in the evolution of today's IoT. Note that the IoT with its underlying machine-to-machine (M2M) communications is useful for enabling the automation of industrial and other machine-centric processes. It is designed to enable communications among machines without relying on any human involvement. Conversely, the Tactile Internet will add a new dimension to HMI involving its inherent *human-in-the-loop* (HITL) nature, thus allowing for a human-centric design approach toward creating novel immersive experiences and extending the capabilities of the human through the Internet, i.e. augmentation rather than automation of the human [28].

In this section, we build on our findings in [28] and aim at tackling the following issues: (i) investigate the potential of leveraging mobile end-user equipment for decentralization via blockchain, (ii) explore how crowdsourcing may be used to decrease the completion time of physical tasks, and (iii) extend the BIoT framework from JC to nudge contract for enabling the nudging of human users in a broader Tactile Internet context.

5.2.1. Ethereum: The DAO

Ethereum made great strides in having its technology accepted as the blockchain standard, when Microsoft Azure started offering it as a service in November 2015. Ethereum was founded by Vitalik Buterin after his request for creating a wider and more general scripting language for the development of decentralized apps (DApps) that are not limited to cryptocurrencies, a capability that Bitcoin lacked, was rejected by the Bitcoin community according to Ethereum white paper "A Next-Generation Smart Contract and Decentralized Application Platform." Ethereum enables new forms of economic organization and distributed models of companies, businesses, and ownership, e.g. self-organized holacracies and member-owned cooperatives. Or as Buterin puts it, although most technologies tend to automate workers on the periphery doing menial tasks, Ethereum automates away the center. For instance, instead of putting the taxi driver out of a job, Ethereum puts Uber out of a job and lets the taxi drivers work with the customer directly (before Uber's self-driving cars will eventually wipe out their jobs). Hence, Ethereum doesn't aim at eliminating jobs so much as it changes the definition of work. In fact, it gave rise to the first DAO built within the Ethereum project. The DAO is an open-source, distributed software that exists "simultaneously nowhere and everywhere," thereby creating a paradigm shift that offers new opportunities to democratize business and enable entrepreneurs of the future to design their own entirely virtual organizations customized to the optimal needs of their mission, vision, and strategy to change the world [27].

A successful example of deploying the DAO concept for automated smart contract operation is *Storj*, which is a decentralized, secure, private, and encrypted cloud storage platform that may be used as an alternative to centralized storage providers such as Dropbox or Google Drive. A DAO may be funded by a group of individuals who cover its basic costs, giving the funders voting rights rather than any kind of ownership or equity shares. This creates an autonomous and transparent system that will continue on the network for as long as it provides a useful service for its customers. DAOs exist as open-source, distributed software that executes smart contracts and works according to specified governance rules and guidelines. Buterin described on the Ethereum Blog the ideal of a DAO as follows: it is an entity that lives on the Internet and exists not only autonomously but also heavily relies on hiring individuals to perform certain tasks that the automation itself cannot do. Unlike AI-based agents that are completely autonomous, a DAO still requires heavy involvement from humans specifically interacting according to a protocol defined by the DAO in order to operate. For illustrating the distinction between a DAO and AI, Figure 5.5 shows a quadrant chart that classifies DAOs, AI, traditional organizations as well as robots, which have been widely deployed in assembly lines among others, with regard to automation and humans involved at their edges and center. We will illustrate a use case of how this particular feature of DAOs (i.e. automation at the center and humans at the edges) can be exploited for decentralizing the Tactile Internet, which represents a promising example of future techno-social systems, below in Section 5.2.2. Toward this end, we also briefly note that according to Buterin a DAO is nonprofit, though one can make money in a DAO, not just by providing investment into the DAO itself but also by participating in its ecosystem, e.g. via membership of human–agent–robot teamwork (HART).

5.2.2. DAO Use Case: Decentralizing Technosocial Systems

In this section, we explore how Ethereum blockchain technologies, in particular the DAO, may be leveraged to decentralize the Tactile Internet as a promising example of future technosocial systems, which at present is yet unclear in many ways how this would work exactly [20].

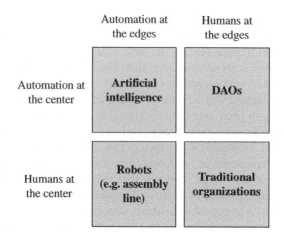

Figure 5.5   DAOs vs. artificial intelligence, traditional organizations, and robots (widely deployed in assembly lines, among others): automation and humans involved at their edges and center. Source: Ethereum Blog/CCBY 4.0).

5.2.2.1. The Tactile Internet: Key Principles

The term "Tactile Internet" was first coined by G. P. Fettweis in 2014. In his seminal paper [29], the Tactile Internet was defined as a new breakthrough enabling unprecedented mobile applications for tactile steering and control of real and virtual objects by requiring a round-trip latency of 1–10 milliseconds. Later in August 2014, ITU-T published the Technology Watch Report "The Tactile Internet," which emphasized that scaling up research in the area of wired and wireless access networks will be essential, ushering in new ideas and concepts to boost access networks' redundancy and diversity to meet the stringent latency as well as carrier-grade reliability requirements of Tactile Internet applications. The Tactile Internet represents one of the most interesting 5G ultra-reliable and low-latency communication (URLLC) applications, which is anticipated to play an important role in 6G as well.

This mandatory end-to-end design approach is fully reflected in the key principles of the reference architecture within the emerging IEEE P1918.1 standards working group, which aims at defining a framework for the Tactile Internet. Among others, the key principles envision to (i) develop a generic Tactile Internet reference architecture, (ii) support local area as well as wide area connectivity through wireless (e.g. cellular, Wi-Fi) or hybrid wireless/wired networking, and (iii) leverage computing resources from cloud variants at the edge of the network. The IEEE P1918.1 standards working group was approved by the IEEE Standards Association in March 2016. The group defines the Tactile Internet as follows: "A network, or a network of networks, for remotely accessing, perceiving, manipulating or controlling real and virtual objects or processes in perceived real time." Some of the key use cases considered in IEEE P1918.1 include teleoperation, haptic communications, immersive VR, and automotive control.

5.2.2.2. FiWi-enhanced Mobile Networks: Spreading Ownership

In [14], we have shown that the 5G URLLC requirements can be achieved by enhancing coverage-centric 4G LTE-A heterogeneous networks (HetNets) with capacity-centric fiber-wireless (FiWi) access networks based on low-cost, data-centric Ethernet passive optical network (EPON) and WLAN technologies in the backhaul and front-end, respectively. Figure 5.6 illustrates the architecture of such FiWi-enhanced mobile networks in greater detail. The common EPON fiber backhaul is shared by a number of ONUs, which may either connect to fixed subscribers or interface with a WLAN MPP or a cellular BS. Some of the resultant ONU-MPPs/ONU-BSs may be equipped with an AI-enhanced MEC server (to be described in more detail shortly). On the end-user side, we consider the following three types of subscribers: conventional mobile users (MUs) as well as pairs of HO and TOR involved in teleoperation, which may be either local or nonlocal.

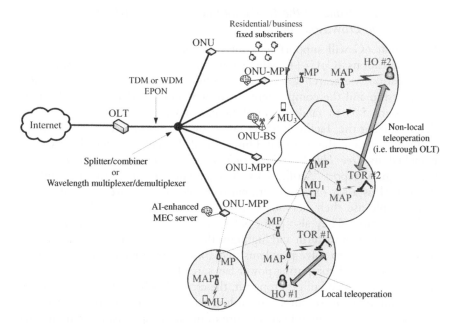

Figure 5.6 Architecture of FiWi-enhanced mobile networks for immersive Tactile Internet applications such as local and nonlocal teleoperation. Source: Maier and Ebrahimzadeh (2019) © 2019 IEEE.

More interestingly, note that in [14], we explored the idea of treating the human as a "member" of a team of intelligent machines rather than keep viewing him as a conventional "user." In addition, we elaborated on the role AI-enhanced agents (e.g. MEC servers) may play in supporting humans in their task coordination between humans and machines. Toward achieving advanced human–machine coordination, we developed a distributed allocation algorithm of computational and physical tasks for fluidly orchestrating hybrid HART coactivities. More specifically, all HART members established through communication a collective self-awareness with the objective of minimizing the task completion time based on the shared use of robots that may be either user- or network-owned. We were particularly interested in the impact of spreading ownership of robots across people whose work they may replace. Our results showed that from a performance perspective (in terms of task completion time) no deterioration occurs if the ownership of robots is shifted entirely from network operators to MUs, though spreading ownership across end-users makes a huge difference in who reaps the benefits from new technologies such as robots. This also applies to blockchain technologies, of course. Recall from above that decentralized ownership is one of the biggest issues to unlock the potential of BIoT for its broader use.

5.2.2.3. Decentralizing the Tactile Internet: AI-Enhanced MEC and Crowdsourcing

Given that 6G will support new service types, e.g. computation oriented communications (CoC) [4], where new smart devices call for distributed computation, we search for synergies between the aforementioned HART membership and the complementary strengths of the DAO, AI, and robots (see Figure 5.5) to facilitate local human–machine coactivity clusters by decentralizing the Tactile Internet [30]. Toward this end, it is important to better understand the merits and limits of AI. Recently, Stanford University launched its *One Hundred-Year Study on Artificial Intelligence (AI100)*. In the inaugural report "Artificial Intelligence and Life in 2030," the authors defined AI as a set of computational technologies that are inspired by the ways people use their nervous systems and bodies to sense, learn, reason, and take action. They also point out that AI will likely replace tasks rather than jobs in the near term and highlight the importance of crowdsourcing of human expertise to solve problems that computers alone cannot solve well. As interconnected computing power has spread around the world and useful platforms have been built on top of it, the crowd has become a demonstrably viable and valuable resource. According to [27], there are many ways for companies that are squarely at the core of modern capitalism to tap into the expertise of uncredentialed and conventionally inexperienced members of the technology-enabled crowd such as the DAO.

AI-Enhanced MEC. According to [4], 6G will go beyond mobile Internet and will be required to support ubiquitous AI services from the core to the end devices. In the following, we explore the potential of leveraging mobile end-user equipment by partially or fully decentralizing MEC. Recall from above that we introduced the use of AI-enhanced MEC servers at the optical-wireless interface of FiWi-enhanced mobile networks. In our considered scenario, we assume that the human–system interface (HSI) at the HO side is equipped with the so-called AI-enabled edge sample forecasting (ESF) module, which is responsible to provide the HO with the predicted samples if the samples are lost or excessively delayed. This as a result helps the HO have a transparent perception of the remote environment. In a BIoT context, these MEC servers have been used as gateways that are required to act as BIoT service agents to release resource-constrained IoT devices from computation-intensive tasks by offloading blockchain transactions onto more powerful edge computing resources. This design constraint can be relaxed in the Tactile Internet, where user equipment (e.g. state-of-the-art smartphones or the aforementioned user-owned robots) is computationally more resourceful than IoT devices and thus may be exploited for decentralization.

In our simulations, we set the WLAN frontend and EPON fiber backhaul network parameters to their default values (see [14] for a more detailed description of the FiWi network parameter setting, system model, and network configuration). We consider $1 \leq N_{Edge} \leq 4$ AI-enhanced MEC servers, each associated with eight end-users, whereof $1 \leq N_{PD} \leq 8$ partially decentralized end-users can flexibly control the amount of offloaded tasks by varying their computation offloading probability. The remaining $8 - N_{PD}$ are fully centralized end-users that rely on edge computing only (i.e. their computation offloading probability equals 1). Note that for $N_{Edge} = 4$, all end-users may offload their computation tasks onto an edge node. Conversely, for $N_{Edge} < 4$, one or more edge nodes are unavailable for computation offloading and their associated end-users fall back on their local computation resources (i.e. fully decentralized). The computational capacity of MEC servers and partially decentralized end-users are set to 1.44 GHz and 185 MHz, respectively.

Figure 5.7 shows the average task completion time vs. computation offloading probability of the partially decentralized end-users for different N_{Edge} and N_{PD}. We observe from Figure 5.7 that for a given N_{Edge}, increasing N_{PD} (i.e. higher level of decentralization) is effective in reducing the average task completion time. Specifically, for $N_{Edge} = 4$, a high decentralization level ($N_{PD} = 8$) allows end-users to experience a reduction of the average task completion time of up to 89.5% by optimally adjusting their computation offloading probability to 0.7. As shown in Figure 5.7, increasing N_{PD} results in not only a decrease of the average execution time but also a decrease of the communication overhead/burden in the wireless front-end of our considered

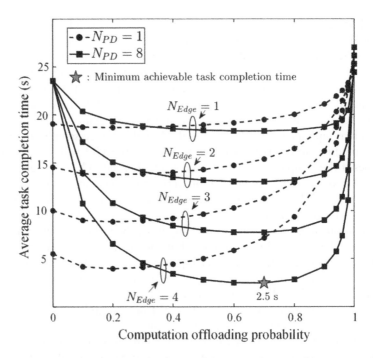

Figure 5.7   Average computational task completion time (in seconds) vs. computation offloading probability for different numbers of partially decentralized end-users (N_{PD}) and AI-enhanced MEC servers (N_{Edge}). Source: Beniiche et al. (2021) © 2021 IEEE.

FiWi network, as a larger number of users shift the computation load from the MEC servers toward their local central processing units (CPUs).

Note that in Figure 5.7, the average task completion time is on the order of seconds, ranging from 2.5 to 25 seconds depending on the computation offloading probability. Hence, given Ethereum's transaction limit of 20 transactions/second, the notoriously low transaction rate of blockchain technologies does not pose a significant challenge to the execution of computational tasks and especially physical tasks carried out by robots in the context of the Tactile Internet, given that the mining process anticipated to be performed by a local CPU of a robot will be offloaded to a nearby MEC server to increase both task completion and transaction confirmation times.

Crowdsourcing.   In [14], we developed a self-aware allocation algorithm of physical tasks for HART-centric task coordination based on the shared use of user- and network-owned robots. By using our AI-enhanced MEC servers as autonomous agents, we showed that delayed force feedback samples coming from TORs may be locally generated and delivered to HOs in close proximity. More specifically, we developed an artificial neural network (ANN)-based forecasting scheme of delayed (or lost) force feedback samples. By delivering the forecast samples to the HO rather than waiting for the delayed ones, we showed that AI-enhanced MEC servers enable HOs to perceive the remote physical task environment in real time at a 1 ms granularity and thus achieve tighter togetherness and improved safety control therein. Note, however, that the performance of the sample forecasting-based teleoperation system heavily relies on the accuracy of the forecast algorithm.

In the following, we explore how crowdsourcing helps decrease the completion time of physical tasks in the event of unreliable forecasting of force feedback samples from TORs. Toward realizing DAO in a decentralized Tactile Internet, Ethereum may be used to establish HO-TOR sessions for remote physical task execution, whereby smart contracts help establish/maintain trusted HART membership and allow each HART member to have global knowledge about all participating HOs, TORs, and MEC servers that act as autonomous agents. We assume that an HO remotely executes a given physical task until X% of the most recently received haptic feedback samples are misforecast. At this point, the HO immediately stops the teleoperation and informs the agent. The agent assigns the interrupted task to a nearby human (e.g. an available HO) in vicinity of the TOR, who then traverses to the task point and finalizes the physical task. The probability of misforecast for a given ESF implementation can be quantified by calculating the long-run average of the ratio of the number of samples that are subject to misforecast to the number of those that are predicted correctly.

Figure 5.8 depicts the average task completion time vs. probability of sample misforecast for different traverse time $T_{traverse}$ of the nearby human and different ratio of human and robot operational capabilities $\frac{f_{human}}{f_{robot}}$, where f_{human} and f_{robot} denote the number of operations per second a human and robot is capable of performing, respectively. We can make several observations from Figure 5.8. Obviously, it is beneficial to select humans with a shorter traverse time, who happen to be closer to the interrupted TOR. We also observe that the ratio $\frac{f_{human}}{f_{robot}}$ has a significant impact on the average task completion time. Clearly, for a ratio of smaller than 1 (i.e. 1/3), the human assistance is less useful since it takes him/her more time to complete the physical task. Conversely, for a ratio of equal to 1 (i.e. 3/3) and especially larger than 1 (i.e. 5/3), crowdsourcing pays off by making use of the superior operational capabilities of the human. Whether humans or robots are better

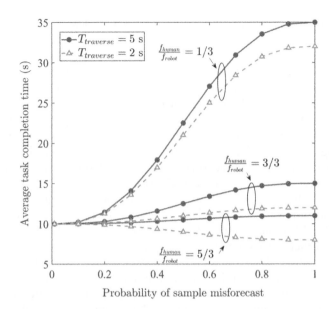

Figure 5.8 Average physical task completion time (in seconds) vs. probability of sample misforecast for different traverse time $T_{traverse} \in \{2,5\}$ seconds of nearby human and different ratio of human and robot operational capabilities $\frac{f_{human}}{f_{robot}}$ (for X=80% fixed). Source: Beniiche et al. (2021) © 2021 IEEE.

suited to perform a physical task certainly depends on its nature. However, for a given physical task, an interesting approach to benefit from the assistance of even uncredentialed and inexperienced crowd members of the DAO may be to enhance the capabilities of humans by means of *nudging*, as explained next.

5.2.3. From AI to Human Intelligence Amplification (IA)

5.2.3.1. Cognitive Assistance

A widely studied approach to increase the usefulness of crowdsourcing has been edge computing, which may be used to guide humans step by step through the physical task execution process by providing them with cognitive assistance. Technically, this could be easily realized by equipping humans with an AR headset (e.g. HoloLens with Wi-Fi connectivity) that receives work-order information in real time from its nearest AI-enhanced

MEC server. In [14], we elaborated on the importance of *shifting the research focus from AI to human intelligence amplification (IA)* by using information technology to enhance human decisions. Note, however, that IA becomes difficult in dynamic task environments of increased uncertainty and real-word situations of great complexity.

5.2.3.2. Hybrid-Augmented Intelligence

Many problems that humans face tend to be of high uncertainty, complexity, and open-ended. To solve such problems, human interaction and participation must be introduced, giving rise to the concept of *hybrid-augmented intelligence* for advanced human–machine collaboration [31]. Hybrid-augmented intelligence is defined as an intelligent model that requires human interaction and allows for addressing problems and requirements that may not be easily trained or classified by machine learning. In general, machine learning is inferior to the human brain in understanding unstructured real-world environments and processing incomplete information and complex spatiotemporal correlation tasks. Hence, machines cannot carry out all the tasks in human society on their own. Instead, AI and human intelligence are better viewed as highly complementary.

According to [31], the Internet provides an immense innovation space for hybrid-augmented intelligence. Specifically, cloud robotics and AR are among the fastest growing commercial applications for enhancing the intelligence of an individual in multirobot collaborative systems. One of the main research topics of hybrid-augmented intelligence is the development of methods that allow machines to learn from not only massive training samples but also human knowledge in order to accomplish highly intelligent tasks via shared intelligence among different robots and humans.

5.2.3.3. Decentralized Self-Organizing Cooperative

A very interesting example to catalyze human and machine intelligence toward a new form of self-organizing artificial general intelligence (AGI) across the Internet is the so-called SingularityNET (https://singularitynet.io), a decentralized AI for the emerging *Global Brain* (see also Chapter 1). One can think of SingularityNet as a decentralized self-organizing cooperative (DSOC), a concept similar to DAO. DSOC is essentially a distributed computing architecture for making new kinds of smart contracts. Entities executing these smart contracts are referred to as agents, which can run in the cloud, on phones, robots, or other embedded devices. Services are offered to any customer via APIs enabled by smart contracts and may require a combination of actions by multiple agents using their *collective intelligence*. In general, there may be multiple agents that can fulfill a given task request in different ways and to different degrees. Each task request to the network requires a unique combination of agents, thus forming a so-called offer network of mutual dependency, where agents make offers to each other to exchange services via offer-request pairs. Whenever someone wants an agent to perform services, a smart contract is signed for this specific task. Toward this end, DSOC aims at leveraging contributions from the broadest possible variety of agents by means of superior discovery mechanisms for finding useful agents and nudging them to become contributors.

5.2.4. Collective Intelligence: Nudging via Smart Contract

Extending on DSOC and the JC introduced at the beginning of this section, we develop a *nudge contract* for enhancing the human capabilities of unskilled crowd members of the DAO. According to Richard H. Thaler, the 2017 Nobel Laureate in Economics, a nudge is defined as any aspect of a choice architecture that alters people's behavior in a predictable way without forbidding any options or significantly changing their economic incentives. Deployed appropriately, nudges can steer people, as opposed to steer objects – real or virtual – as done in the conventional Tactile Internet, to make better choices and positively influence the behavior of crowds of all types.

Our nudge contract aims at completing interrupted physical tasks by learning from a skilled DAO member with the objective of minimizing the learning loss, which denotes the difference between the achievable and optimum task execution times. Given the reward enabled by the nudge contract and associated with each skill transferred, a remote skilled DAO member submits a hash address of the learning instructions to an unskilled human/robot. The hash address is stored on the blockchain, whereby the corresponding data of the learning instructions may be stored on a remote decentralized storage server, e.g. Inter-Planetary File System (IPFS). An unskilled human/robot can retrieve the learning instructions using the corresponding hash address. The ability to learn a given subtask[2] is characterized by the subtask learning probability. The learning process is accomplished if each subtask is learned successfully from a skilled DAO member, who in turn is rewarded via a smart contract (see Algorithm 5.1 for details).

---

[2]We assume that the incoming physical tasks are decomposable, meaning that they can be broken into a number of subtasks. An example of such tasks can be the part assemblage in an industry automation scenario.

Algorithm 5.1: Nudge Contract

---

1 **Given:** Set $U = \{h_1, h_2, ..., h_n\}$ of n DAO members, capability vector $\mathbf{C} = [c_1, c_2, ..., c_n]$, distance vector $\mathbf{D} = [d_1, d_2, ..., d_n]$, interrupted task $\mathbf{T}$, required number D of actions to execute the interrupted task, interrupted robot r_0, capability requirement c_0 of the interrupted task

2 Decompose the given interrupted task $\mathbf{T}$ into N_{sub} subtasks

3 **for** $i = 1$ *to* n **do**

4 | **if** $c_i \geq c_0$ **then**

5 | | $S \leftarrow h_i$

6 | **end**

7 **end**

8 $h^* \leftarrow \arg\min_{d_i} \{S\}$

9 Create a secure blockchain transaction between h^* and interrupted robot r_0

10 Send the learning instructions from h^* to r_0 through the established transaction

11 Use the multiarm bandit selection strategy in [32] to help the robot learn the given set of subtasks

12 **if** *all N_{sub} subtasks are learned successfully* **then**

13 | learning process is successfully accomplished

14 | r_0 can execute the interrupted task $\mathbf{T}$ with the capability of h^*

15 | **else**

16 | Learning process is failed

17 | DAO member h^* traverses to the interruption point to execute the task $\mathbf{T}$

18 **end**

19 Reward the skilled DAO member h^* via blockchain smart contract

---

Figure 5.9 shows the performance of our nudge contract for 50 DAO crowd members, whose ratio $\frac{f_{human}}{f_{robot}}$ is randomly chosen from $\{1/3, 3/3, 5/3\}$. We observe that for a given subtask learning probability, decreasing the number N_{sub} of subtasks helps reduce the learning loss, thus indicating the importance of a proper task decomposition method. More specifically, over-decomposition of the given task will result in performance deterioration unless a suitable learning mechanism is adopted to increase the subtask learning probability.

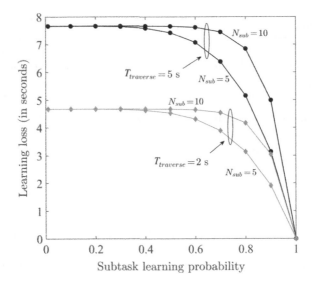

Figure 5.9 Learning loss (in seconds) vs. subtask learning probability for different number N_{sub} of subtasks and traverse time $t_{traverse}$ [30]. Source: Beniiche et al. (2021) © 2021 IEEE.

References

1. M. Weiser. The computer for the 21st century. *Scientific American*, 265(3):94–104, September 1991.
2. I. Ahmad, T. Kumar, M. Liyanage, M. Ylianttila, T. Koskela, T. Braysy, A. Anttonen, V. Pentikinen, J. Soininen, and J. Huusko. Towards gadget-free internet services: a roadmap of the naked world. *Elsevier Telematics and Informatics*, 35(1):82–92, April 2018.
3. E. Bastug, M. Bennis, M. Médard, and M. Debbah. Toward interconnected virtual reality: opportunities, challenges, and enablers. *IEEE Communications Magazine*, 55(6):110–117, June 2017.
4. K. B. Letaief, W. Chen, Y. Shi, J. Zhang, and Y. A. Zhang. The roadmap to 6G: AI empowered wireless networks. *IEEE Communications Magazine*, 57(8):84–90, August 2019.
5. E. C. Strinati, S. Barbarossa, J. L. Gonzalez-Jimenez, D. Kténas, N. Cassiau, L. Maret, and C. Dehos. 6G: The next frontier: from holographic messaging to artificial intelligence using subterahertz and visible light communication. *IEEE Vehicular Technology Magazine*, 14(3):42–50, September 2019.
6. B. Zong, C. Fan, X. Wang, X. Duan, B. Wang, and J. Wang. 6G technologies: key drivers, core requirements, system architectures, and enabling technologies. *IEEE Vehicular Technology Magazine*, 14(3):18–27, September 2019.

7. K. David and H. Berndt. 6G vision and requirements: is there any need for beyond 5G? *IEEE Vehicular Technology Magazine*, 13(3):72–80, September 2018.
8. W. Saad, M. Bennis, and M. Chen. A vision of 6G wireless systems: applications, trends, technologies, and open research problems. *IEEE Network*, 34(3):134–142, May/June 2020.
9. ABI Research and Qualcomm. Augmented and Virtual Reality: the First Wave of 5G Killer Apps. *White Paper*, February 2017.
10. Å. Fast-Berglund, L. Gong, and D. Li. Testing and validating extended reality (xR) technologies in manufacturing. *Procedia Manufacturing*, 25:31–38, May 2018.
11. Nissan Newsroom. Nissan Unveils Invisible-to-Visible Technology Concept at CES: Future Connected-car Technology Merges Real and Virtual Worlds to Help Drivers 'See the Invisible', January 2019. [Online; Accessed on 2022-06-09]. Available: https://global.nissannews.com/ja-JP/releases/nissan-unveils-invisible-to-visible-technology-concept-at-ces.
12. S. Haddadin, L. Johannsmeier, and F. Díaz Ledezma. Tactile robots as a central embodiment of the tactile internet. *Proceedings of the IEEE*, 107(2):471–487, February 2019.
13. G. Dublon and J. A. Paradiso. Extra sensory perception. *Scientific American*, 311(1):36–41, July 2014.
14. M. Maier and A. Ebrahimzadeh. Towards immersive tactile internet experiences: low-latency FiWi enhanced mobile networks with edge intelligence [invited]. *IEEE/OSA Journal of Optical Communications and Networking, Special Issue on Latency in Edge Optical Networks*, 11(4):B10–B25, April 2019.
15. D. Bohm. *Wholeness and the Implicate Order*. Routledge, March 2002.
16. A. S. Cacciapuoti, M. Caleffi, F. Tafuri, F. S. Cataliotti, S. Gherardini, and G. Bianchi. Quantum internet: networking challenges in distributed quantum computing. *IEEE Network*, 34(1):137–143, January/February 2020.
17. A. S. Cacciapuoti, M. Caleffi, R. Van Meter, and L. Hanzo. When entanglement meets classical communications: quantum teleportation for the quantum internet (invited paper). *IEEE Transactions on Communications*, 68(6):3808–3833, June 2020.
18. D. Buonomano. *Your Brain Is a Time Machine: The Neuroscience and Physics of Time*. W. W. Norton, April 2017.
19. D. Tapscott and A. Tapscott. *Blockchain Revolution: How the Technology Behind Bitcoin Is Changing Money, Business, and the World*. Portfolio, Toronto, ON, Canada, May 2016.
20. R. Beck. Beyond bitcoin: the rise of blockchain world. *IEEE Computer*, 51(2):54–58, February 2018.
21. O. Novo. Blockchain meets IoT: an architecture for scalable access management in IoT. *IEEE Internet of Things Journal*, 5(2):1184–1195, April 2018.
22. T. M. Fernández-Caramés and P. Fraga-Lamas. A review on the use of blockchain for the Internet of Things. *IEEE Access*, 6:32979–33001, 2018.

23. J. Pan, J. Wang, A. Hester, I. Alqerm, Y. Liu, and Y. Zhao. EdgeChain: An edge-IoT framework and prototype based on blockchain and smart contracts. *IEEE Internet of Things Journal*, 6(3):4719–4732, June 2019.

24. Y. Zhang, S. Kasahara, Y. Shen, X. Jiang, and J. Wan. Smart contract-based access control for the Internet of Things. *IEEE Internet of Things Journal*, 6(2):1594–1605, April 2019.

25. C.-F. Liao, S.-W. Bao, C.-J. Cheng, and K. Chen. On Design Issues and Architectural Styles for Blockchain-driven IoT services. In *Proc., IEEE International Conference on Consumer Electronics - Taiwan (ICCE-TW)*, pages 351–352, June 2017.

26. P. Urien. Towards Secure Elements For Trusted Transactions in Blockchain and Blockchain IoT (BIoT) Platforms. In *Proc., Fourth International Conference on Mobile and Secure Services (MobiSecServ)*, pages 1–5, February 2018.

27. A. McAfee and E. Brynjolfsson. *Machine, Platform, Crowd: Harnessing Our Digital Future*. W. W. Norton, June 2017.

28. M. Maier, A. Ebrahimzadeh, and M. Chowdhury. The tactile internet: automation or augmentation of the human? *IEEE Access*, 6:41607–41618, 2018.

29. G. P. Fettweis. The tactile internet: applications and challenges. *IEEE Vehicular Technology Magazine*, 9(1):64–70, March 2014.

30. A. Beniiche, A. Ebrahimzadeh, and M. Maier. The way of the DAO: toward decentralizing the tactile internet. *IEEE Network*, 35(4):190–197, July/August 2021.

31. N. Zheng, Z. Liu, P. Ren, Y. Ma, S. Chen, S. Yu, J. Xue, B. Chen, and F. Wang. Hybrid-augmented intelligence: collaboration and cognition. *Springer Frontiers of Information Technology & Electronic Engineering*, 18(2):153–179, February 2017.

32. S. McGuire, P. M. Furlong, C. Heckman, S. Julier, D. Szafir, and N. Ahmed. Failure is not an option: policy learning for adaptive recovery in space operations. *IEEE Robotics and Automation Letters*, 3(3):1639–1646, July 2018.

CHAPTER 6

Web3 and Token Engineering

"The day is not far off when the economic problem will take the back seat where it belongs, and the arena of the heart and the head will be occupied or reoccupied, by our real problems - the problems of life and of human relations, of creation and behavior and religion."

JOHN MAYNARD KEYNES
Britain's Most Famous 20th-Century Economist (1883–1946)

6.1. Emerging Web3 Token Economy

In Chapter 1, we have briefly described the evolution of the Internet economy from the original read-only Web1 information economy and today's read-write Web2 platform economy to the emerging *read-write-execute Web3* token economy based on decentralized blockchain technologies, as illustrated in Figure 1.2. The Web3 will enable the token economy where anyone's contribution is compensated with a token, as opposed to today's Web2 platform economy that exploits users' free data for targeted advertising.

The exploration of tokens, in particular different types and roles, is still in the very early stages. The understanding of how to apply these tokens is still vague, especially the important problem of *token engineering*, which is an emerging term defined as the theory, practice, and tools to analyze, design, and verify tokenized ecosystems. One of the most comprehensive and up-to-date references on the token economy is Shermin Voshmgir's second edition of her widely cited book titled "Token Economy: How the Web3 reinvents the Internet," which attempts to summarize the existing knowledge about blockchain networks as the backbone of the Web3 and contextualizes its socioeconomic implications [1]. In the following, we will outline Voshmgir's main findings and insights in a comprehensive yet comprehensible manner.

6G and Onward to Next G: The Road to the Multiverse, First Edition. Martin Maier.
© 2023 The Institute of Electrical and Electronics Engineers, Inc.
Published 2023 by John Wiley & Sons, Inc.

6.2. Backend Revolution

In the read-only Web1, Tim Berners Lee introduced the Hypertext Markup Language (HTML), the standard markup language for documents designed to be displayed in a web browser, which allowed surfing the Internet following links instead of using command-line interfaces. As a result, the Internet became more usable and anyone could now use it.

With the read–write Web2, the Internet became more mature. Applications could be used to read and write simultaneously. This revolutionized social and economic interactions, bringing producers and consumers of information, goods, and services closer together, but always with a middleman: a platform acting as a trusted intermediary between two people who do not know or trust each other.

In the emerging Web3, the read–write frontend remains the same, but the data structures in the backend change. Specifically, smart contracts are used to replace entire back offices and automate them via self-executing software. Conversely, at the network periphery, anyone can participate in verifying blockchain transactions and be compensated for their contribution with a network token. Agreements are executed on the fly and P2P with smart contracts. A smart contract is a self-verifying, self-executing, tamper-resistant piece of software that provides lower transactions costs and higher transparency. Smart contracts can map legal obligations into an automated process. If implemented correctly, they can provide a greater degree of contractual security at lower costs than current legal systems. Web3 applications need a connection to a blockchain or distributed ledger, which is managed by a special application called *wallet*.

6.3. Technology-Enabled Social Organisms

Techno-social systems handle control of transactions through technical systems that can be autonomous. According to [2], it is unclear in many ways how this would work exactly, but through blockchain technology we have the chance to experiment with secure, decentralized systems, which could enable new social models that go well beyond the economy.

Web3 networks create complex technology-enabled social organisms that require an iterative social *governance* process of finding consensus about policy upgrades. Governance is the term that is colloquially used by many to describe the social consensus process over protocol evolution. This process can be conducted either off-chain or on-chain.

A challenge with current proposals for on-chain governance is that they are plutocratic, which means that protocol upgrades are decided proportional to one's token holdings. Token holders with more tokens would therefore have more voting power than smaller token holders. This is a considerable design question, given that token distribution is often disproportionately uneven. For instance, in May 2016, from a total of 11,000 investors, the top 100 holders held over 46% of all The DAO tokens. In light of such plutocratic voting mechanisms, using the term decentralization could be perceived as contradictory. Off-chain governance, on the other hand, is relatively centralized and excludes many small token holders, especially those who lack the technical knowledge or financial power to assess network decisions adequately. It is still unclear what the right balance between on-chain vs. off-chain governance could look like.

6.4. Tokens

The term cryptocurrency is not ideal, since many of these new assets were never issued with the intention to represent money in the first place. The term *token* is becoming more widespread since it is more generic, encompassing all token types rather than only asset-backed tokens. Tokens might be the killer application of blockchain networks. These tokens are often issued with just a few lines of code in the form of a smart contract that is collectively managed by a blockchain network or a similar distributed ledger. All nodes in the network have the same information about who owns which tokens and the transfer of those tokens is collectively managed.

Tokens are only accessible with a dedicated wallet software that communicates with the blockchain network and manages the public–private key pair related to the blockchain address. Only the person who has the private key for that address can access the respective tokens. This person can, therefore, be regarded as the owner or custodian of that token. If the token represents an asset, the owner can initiate transfer of the token by signing with her private key. If the token represents an access right to something somebody else owns, the owner of that token can also initiate access by signing with her private key. The same applies to tokens that represent voting rights.

6.4.1. Tokenization: Creation of Tokenized Digital Twins

Tokens are instrumental in enabling the important process of *tokenization*. The tokenization of an existing asset refers to the process of creating a *tokenized digital twin* for any physical object or financial asset. Once tokenized, every kind of value can be managed as a digital asset, whose unit of account is a dedicated token. The tokens can be minted by any individual or organization that defines the set of rules governing them.

6.4.2. Token Contract

A *token contract* is a special type of smart contract that defines a bundle of conditional rights assigned to the token holder. They are right management tools that can represent any existing digital or physical asset, or access rights to assets someone else owns. Tokens can represent anything from a store of value to a set of permissions in the physical, digital, and legal world. They facilitate collaboration across markets and jurisdictions and allow more transparent, efficient, and fair interactions between market participants at low costs. Tokens can also incentivize an autonomous group of people to individually contribute to a collective goal. These tokens are created upon proof of a certain behavior.

The ability to deploy tokens at a low cost relatively effortlessly on a public infrastructure is a game changer, because it makes it economically feasible to represent many types of assets and access rights in a digital way that might not have been feasible before. The use of tokens could also enable completely new use cases, business models, and asset types that were not economically feasible before, and potentially enable completely new value-creation models. Note that tokens are not a new thing and have existed long before the emergence of blockchain networks. Traditionally, tokens can represent any form of economic value or access right. In cognitive psychology, tokens have been used as a positive reinforcement method of incentivizing desirable behavior in patients. Cognitive psychology uses reward tokens as a medium of exchange.

The first blockchain tokens were the native tokens of public and permissionless blockchain networks. These native tokens are also referred to as *protocol tokens*. With the advent of Ethereum, however, tokens have moved up the technology stack and can now be issued on the application layer, giving rise to so-called *application tokens*. The Ethereum network has one protocol token, i.e. ETH, and a whole economy of application tokens running on top of the network. ERC-20 and other Ethereum token standards allow the creation of application tokens with a smart contract (see also token contract mentioned above).

6.4.3. Fungible Tokens vs. Non-Fungible Tokens (NFTs)

Importantly, we are still in the very early stages of exploring different roles and types of tokens. In general, there are two types of token: (i) fungible tokens and (ii) non-fungible tokens, also known as NFTs. *Fungibility* refers to the interchangeability of a unit of an asset with other units of the same asset. For illustration, Table 6.1 summarizes the main differences between fungible and non-fungible tokens with regard to interchangeability and divisibility.

Another question when designing a token is whether the token has an *expiration date*. Any fungible token might be programmed in a way that

Table 6.1   Tokens: taxonomy and attributes.

Fungible tokens	Non-fungible tokens (NFTs)
Identical	*Unique*
Tokens of the same type are identical to one another.	Each token is unique and differs from all other tokens of the same type.
They have identical attributes.	They have unique attributes.
Interchangeable	*Non-interchangeable*
A token can be interchanged for another with the same value. A 20$ bill can be replaced with a combination of other bills and coins that amount to the same value.	NFTs cannot be replaced with tokens of the same type as they represent unique values or access rights.
	A university degree or driver's license would be non-interchangeable.
Divisible	*Non-divisible*
Fungible assets are divisible into smaller amounts. It is irrelevant which and how many units one uses as long as it adds up to the same value.	Tokens that are tied to one's identity, like certificates and degrees, are not divisible.
	It does not make sense to have a fraction of a degree or driver's license.

Source: *Voshmgir [1].*

it expires after a certain date in order to prevent hoarding of the tokens. Practically speaking, the token would expire; technically speaking, the token would change state. In the past, some regional currencies, like the Austrian "Wörgl Schwundgeld" in the 1930s, experimented with an inbuilt deflation of their currency to prevent hoarding and inflation. The currency was introduced as a parallel currency that could only be spent in the region of Wörgl. By losing 1% of its value each month (i.e. negative interest rate), individual spending was encouraged while saving was disincentivized. This measure was introduced to successfully help with both unemployment numbers and infrastructure investment.[1]

---

[1]The idea of money with an expiration date has a long history, e.g. Silvio Gesell's "free money" (or *Freigold*) – "free" because it would be freed from hoarding and encourage bankers to lend money without charging interest. To avoid expiration, bills would have to be periodically stamped for a fee. With no new stamp, they would become worthless. Only by spending or investing money would you be able to avoid stamp fees. John Maynard Keynes dedicated five pages to Gesell in a concluding chapter of his magnum opus *The General Theory of Employment, Interest and Money*. While critiquing some of Gesell's overall theory, Keynes concluded, "The idea behind stamped money is sound." Another noteworthy example of a new kind of money is the so-called *Edison-Ford commodity money* envisioned by Thomas A. Edison and Henry Ford to aid farmers and to abolish speculation in farm products by divorcing agriculture from the banking system.

6.4.4. Purpose-Driven Tokens

6.4.4.1. Public Goods and Externalities

Purpose-driven tokens incentivize individual behavior to contribute to a collective goal. This collective goal might be a *public good* or the reduction of *negative externalities* to a common good. Purpose-driven tokens introduce a new form of collective value creation in the absence of traditional intermediaries. They provide an alternative to the conventional economic system, which predominantly incentivizes individual value creation in the form of private goods. In addition to maximizing one's personal profit, purpose-driven tokens aim at incentivizing behavior toward a collective purpose beyond individual profit. Thus, purpose-driven tokens provide an operating system for a new type of economy.

Externalities in economics refer to the costs (i.e. negative externalities) or benefits (i.e. positive externalities) that affect a third person or community, who did not choose to participate in the economic transaction and incur that cost or benefit. *Positive externalities* can arise if, for example, two neighboring farmers have positive ecological effects on each other. Note, however, that even though the collective production of public goods can result in positive externalities, it does not necessarily exclude other negative externalities. If not well designed, purpose-driven tokens can have positive and negative externalities. While proof-of-work (PoW) is an essential blockchain mechanism for the maintenance of a public good, the act of mining itself is energy intense, producing negative externalities to society.

Blockchain networks like Ethereum took the idea of collective value creation to the next level by providing a public infrastructure for creating an application token with only a few lines of code. With these application tokens, we can create completely new types of economies with a simple smart contract that runs on a public and verifiable infrastructure. These tokens are an easily programmable vehicle to model individual decision-making processes into a smart contract. In principle, any purpose can be incentivized. As a consequence, Ethereum and similar blockchain networks provide a public good similar to the ones governments usually provide to their citizens. However, as opposed to state-controlled public goods, blockchain networks have distributed upkeep, development, and control.

Depending on the design of the token, rewards could be exchanged for some other services provided by the organization issuing purpose-driven tokens. They can vary greatly from project to project. For instance, the purpose-driven tokens "Sweatcoin" or "Changers" incentivize riding a bike, walking, or using public transportation instead of using a car. Other projects incentivize the production or consumption of renewable energies such as "Solar Coin," "Electric Chain," and "Sun Exchange." One could also be incentivized for planting trees or cleaning a beach, reduction of food waste, and many more. Some illustrative examples of such purpose-driven tokens include "Plastic Bank," "Earth Token," or "Eco Coin."

Conventional economic systems predominantly incentivize individual value creation: private actors to extract rent from nature or from the workforce, and transform this into products, often externalizing costs to society, while internalizing private profits. However, this new and collective value creation that purpose-driven tokens introduce will likely need much more research and development, and a long phase of trial and error, before we can better understand the potential of incentivizing contributions to a public good. Operational use cases are still limited.

Apart from the aforementioned built-in expiry date, tokens can also be programmed to have *limited transferability* such that they can only be exchanged for local products and services of the community, therefore never leaving the internal system and being exchanged for fiat money, but still useful in the internal economy of a network by serving as a kind of community currency. Toward this end, the study of economics, public choice theory, theory of public goods, and behavioral sciences will be essential for a better understanding, and as a result, also a better design and engineering process of purpose-driven tokens.

6.4.4.2. Club Goods and Tech-Driven Public Goods

Of particular importance for the proper design of purpose-driven tokens is hereby the term public goods. In economics, public goods refer to goods that any individual can use without paying for them (i.e. non-excludable) and where use by one individual does not reduce the availability to others (i.e. non-rivalrous). With physical goods, consumption by one person prevents that of another. The case is different for digital goods, which can easily be copied and distributed. Public goods can be provided by a government or be available in nature. Examples thereof include knowledge, the Internet, and certain natural resources such as air. Public goods tend to be subject to *free-rider problems*, giving rise to the well-known tragedy of the commons. If a certain threshold of people and institutions decide to free-ride, the market will fail to provide a good or service for which there is a need. Many free-rider and tragedy of the commons problems need to be anticipated when designing the token governance mechanisms discussed above. If public goods become subject to restrictions, they become *club goods* (or private goods). Exclusion mechanisms might be in the form of membership, among others. Club goods represent artificially scarce goods.

Purpose-driven tokens can be seen as a new form of creating *tech-driven public goods*. They can be programmed to maintain or restore a common good and could possibly resolve many tragedy-of-the-commons problems society faces today. More specifically, purpose-driven tokens could provide a mechanism for *nudging* individuals via positive incentives to collectively contribute to the creation of public goods and/or reduction of negative externalities of a common good. Nudging suggests building on the assumption of bounded rationality and suggests that individuals can be supported in their decision-making process, which is impacted by other factors than economic rationality. Psychological, emotional, cultural, cognitive, and social factors are also taken into account, with the conclusion that people make over 90% of their decisions based on mental shortcuts or rules of thumb. Especially under pressure and in situations of high uncertainty, humans tend to rely on anecdotal evidence and stereotypes to help them understand and respond to events more quickly. It is assumed that the rationality of individuals and institutions is bounded by time and cognitive limitations, and that good enough solutions are preferred over perfect solutions. *Behavioral economics* builds on the learnings of cognitive psychology, a field of psychology that studies mental processes. Alternative economic theories, such as behavioral economics, are based on the assumption that individual action is more complex. Behavioral economics studies why market actors are economically irrational and how others can profit from such (predictable) irrationality.

Critics, however, argue that nudging equals psychological manipulation and social engineering. Purpose-driven tokens can also be used to nudge or steer individuals toward certain actions, e.g. reducing CO_2 emissions. However, any type of governance system is steering collective action and per definition aims at social engineering. Behavioral economics and methods like nudging can therefore provide important tools when designing the governance rules of purpose-driven tokens as a means to provide public goods. There is much we can learn, and not only from the ethics of behavioral economics. There is an entire class of games in the network science literature called network formation games. As the field of purpose-driven tokens matures, it is likely that *behavioral game theory* will find its way into the modeling of purpose-driven tokens (to be further explored in the context of the classical trust game of behavioral economics in Chapter 7). Related disciplines like mechanism design have also developed ethical principles that can be relevant to token engineering, as discussed next.

6.5. Mechanism Design and Token Engineering

Mechanism design is a subfield of economics that deals with the question of how to design a game that incentivizes everyone to contribute to a collective goal.

It is also referred to as "reverse game theory," since it starts at the end of the game (i.e. its desirable outcome), then goes backward when designing the mechanism. In 2007, the Nobel Prize in Economics was awarded to Leonid Hurwicz, Eric Maskin, and Roger Myerson for their contributions to mechanism design theory.

According to Eric Maskin, we can think of mechanism design as the *engineering part of economic theory*. Most of the time in economics, we look at existing economic institutions and try to explain or predict the outcomes that those institutions generate. This is called the positive or predictive part of economics. Perhaps 80–90% of economists do positive economics. In mechanism design, we do just the opposite by starting with the outcomes. We identify the outcomes we would like to have. And then we work backwards to see whether we can engineer institutions (mechanisms) that will lead to those outcomes. This is the normative or prescriptive part of economics. It may not be what most economists do, but it is extremely important too.[2]

Purpose-driven tokens need purpose-oriented mechanisms. Token mechanism design, also referred to as *token engineering*, is an emerging field. However, best practices are scarce and many of the existing use cases have considerable design flaws. Real-world use cases are complex and require data feeds from the outside world, such as blockchain oracles (to be explained shortly in Chapter 7). In order to be able to adequately address issues like the tragedy of the commons and free-rider problems, we need a much more nuanced mechanism design of purpose-driven tokens.

Token engineering provides a more interdisciplinary approach that comprises the theory, practice, and tools to analyze, design, and verify tokenized ecosystems. While a lot of tokens have been issued through token sales over the last few years – mostly for fundraising purposes – most of these tokens lack proper functionality and mechanism design. So far, there has been little overlap with the academic community studying this field and the developers of many purpose-driven tokens. Among others, open research challenges in the emerging field of token mechanism design include the design of a bottom-up token engineering framework to enable future state-of-the-art design of tokenized ecosystems. In Chapter 8, we will elaborate on how a suitable bottom-up multilayer token engineering framework for Society 5.0 may look like by leveraging on not only emerging Cyber-Physical-Social System (CPSS) but also advanced blockchain technologies such as purpose-driven tokens and blockchain oracles.

---

[2] Eric Maskin, "Introduction to mechanism design and implementation," Transnational Corporations Review, April 2019. Available online: https://doi.org/10.1080/19186444.2019.1591087 (accessed on 22 February 2022).

References

1. S. Voshmgir. *Token Economy: How the Web3 Reinvents the Internet* (Second Edition. BlockchainHub, Berlin, Germany, June 2020.
2. R. Beck. Beyond bitcoin: the rise of blockchain world. *IEEE Computer*, 51(2):54–58, February 2018.

CHAPTER 7

From Robonomics to Tokenomics

"There is one word which may serve as a rule of practice for all one's life - reciprocity."

CONFUCIUS
(551 BC–479 BC)

7.1. Robonomics

This chapter focuses on the emerging field of *robonomics*, which studies the sociotechnical impact of blockchain technologies on social human–robot interaction (sHRI), behavioral economics, behavioral game theory, and cryptocurrencies (both coins and tokens) for the social integration of robots into human society. Robonomics involves persuasive robotics, whereby a physical or virtual robotic agent is used as enforcer or supervisor of human behavior modification via psychological rewards in addition to tangible rewards [1].

Advanced blockchain technologies such as *oracles* enable the *on-chaining* of blockchain-external off-chain information stemming from human users. In doing so, they leverage on human intelligence rather than machine learning only. Toward this end, we investigate the widely studied trust game of behavioral economics in a blockchain context, paying close attention to the importance of developing efficient cooperation and coordination technologies. After identifying open research challenges of blockchain-enabled implementations of the trust game, we first develop a smart contract that replaces the experimenter in the middle between trustor and trustee and demonstrate experimentally that a social efficiency of up to 100% can be

6G and Onward to Next G: The Road to the Multiverse, First Edition. Martin Maier.
© 2023 The Institute of Electrical and Electronics Engineers, Inc.
Published 2023 by John Wiley & Sons, Inc.

achieved by using deposits to enhance both trust and trustworthiness. We then present an on-chaining oracle architecture for a networked N-player trust game that involves a third type of human agents called observers, who track the players' investment and reciprocity. The presence of third-party reward and penalty decisions helps raise the average normalized reciprocity above 80%, even without requiring any deposit. Further, we experimentally demonstrate that mixed logical-affective persuasive strategies for social robots improve the trustees' trustworthiness and reciprocity significantly [2].

Finally, the chapter explains the anticipated paradigm shift from conventional monetary to future nonmonetary economies based on technologies that can measure activities toward human co-becoming that have no monetary value. Specifically, the chapter elaborates on the shift from conventional monetary economics to nonmonetary *tokenomics* enabled by tokenization in different value-based scenarios.

7.2. Blockchain Oracles and On-Chaining

A major limitation of the conventional blockchain is its inability to interact with the "outside world" since smart contracts can only operate on data that is on the blockchain. In the emerging blockchain-based Internet of things (BIoT), sensors are typically deployed to bring sensor measurement data onto the blockchain [3]. Advanced blockchain technologies enable the on-chaining of blockchain-external off-chain information stemming also from real users, apart from sensors and other data sources only, thus leveraging also on human intelligence rather than machine learning only. To overcome this limitation, smart contracts may make use of oracles, which are trusted decentralized blockchain entities whose primary task is to collect off-chain information and bring it onto the blockchain as trustworthy input data to smart contracts. Several decentralized oracle systems exist that rely on *voting-based games*, e.g. ASTRAEA [4].

Blockchain-external data sources imply the risk that the on-chained data may be unreliable, maliciously modified, or untruthfully reported. Typically, various game-theoretical mechanisms are used to incentivize truthful provisioining of data. According to [5], however, those approaches address only partial aspects of the larger challenge of assuring trustworthiness in data on-chaining systems. A key property of trustworthy data on-chaining systems is truthfulness, which means that no execution of blockchain state transition is caused by untruthful data provisioning, but instead data is always provisioned in a well-intended way. The challenge that derives from truthfulness is the building of incentive compatible systems, where participants are assumed to act as rational self-interest driven *homini oeconomici*, whose primary goal is to

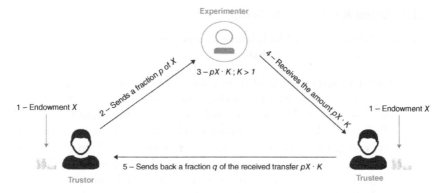

Figure 7.1 Classical trust game involving two human players (trustor and trustee) and one experimenter in the middle.
Source: Beniiche et al. (2021) © 2021 IEEE.

maximize their individual utility via monetary rewards and penalties for their actions and behavior.

In the following, we focus on the *trust game* widely studied in behavioral economics. The trust game hasn't been investigated in a blockchain context yet, though it allows for a more systematic study of not only trust and trustworthiness but also reciprocity between human actors [6]. The classical trust game involves only two human players referred to as trustor and trustee, who are paired anonymously and are both endowed with a certain amount X of monetary units. Figure 7.1 illustrates the sequential exchange between trustor and trustee. The trustor can transfer a fraction $0 \leq p \leq 1$ of her endowment to the trustee. The experimenter then multiplies this amount by a factor $K > 1$, e.g. doubled or tripled. The trustee can transfer a fraction $0 \leq q \leq 1$ of the received amount directly back to the trustor without going through the experimenter. Note that the trust game captures any generic economic exchange between two actors. According to [7], the trust game will remain an important instrument for the study of social capital and its relation to economic growth for many years to come, whereby research on efficient cooperation and coordination technologies will be of particular interest.

7.3. Blockchain-Enabled Trust Game

In this section, we first identify open research challenges, followed by a blockchain-enabled baseline implementation of the trust game for benchmark comparison via experiments.

7.3.1. Open Research Challenges

The use of decentralized blockchain technologies for the trust game should tackle the following research challenges:

- **Social Efficiency**: Recall from above that the trust game allows the study of social capital for achieving economic growth. Toward this end, the closely related term *social efficiency* plays an important role. Social efficiency is defined as the optimal distribution of resources in society, taking into account so-called externalities as well. In general, an externality is the cost or benefit that affects third parties other than the voluntary exchange between a pair of producer and consumer. We will study the impact of externalities below, when we extend the classical trust game to multiplayer games.

  We measure social efficiency as the ratio of the achieved total payoff of both trustor and trustee and the maximum achievable total payoff, which is equal to $X(K + 1)$. A social efficiency of 100% is achieved if the trustor sends her full endowment X (i.e. $p = 1$), which is then multiplied by K, and the trustee reciprocates by sending back the received amount XK fully or in part, translating into a total payoff of $q \cdot XK + (1 - q) \cdot XK + X = X(K + 1)$. Note that maximizing the total payoffs requires to set $p = 1$ for a given value of K, though q may be set to any arbitrary value. The parameter q, however, plays an important role in controlling the (equal or unequal) distribution of the total payoffs between trustor and trustee, as discussed in more detail shortly. Conversely, if the trustor decides to send nothing (i.e. $p = 0$) due to the lack of trust (on the trustor's side) and/or lack of trustworthiness (on the trustee's side), both are left with their endowment X and the social efficiency equals $2X/X(K + 1) = 2/(K + 1)$. How to improve social efficiency in an equitable fashion in a blockchain-enabled trust game is an important research challenge.

- **Trust and Trustworthiness in N-Player Trust Game**: In the past, games of trust have been limited to two players. Abbass et al. [8] introduced a new N-player trust game that generalizes the concept of trust, which is normally modeled as a sequential two-player game to a population of multiple players that can play the game concurrently. According to [8], evolutionary game theory shows that a society with no untrustworthy individuals would yield maximum wealth to both the society as a whole and the individual in the long run. However, when the initial population consists of even the slightest number of untrustworthy individuals, the society converges to zero trustors. The proposed N-player trust game shows that the promotion of trust is an uneasy task, despite the fact that a combination of trustors and trustworthy trustees is the most rational and optimal social state.

It's important to note that the N-player trust game in [8] was played in an unstructured environment, i.e. the population was not structured in any specific spatial topology or social network. Chica et al. [9] investigated whether a *networked* version of the N-player trust game would promote higher levels of trust and global net wealth (i.e. total payoffs) in the population than that of an unstructured population. To do so, players were mapped to a spatial network structure, which restricts their interactions and cooperation to local neighborhoods. Unlike [8], where the existence of a single untrustworthy individual would eliminate trust completely and lead to zero global net wealth, Chica et al. [9] discovered the importance of establishing network structures for promoting trust and global net wealth in the N-player trust game in that trust can be promoted despite a substantial number of untrustworthy individuals in the initial population. Clearly, the development of appropriate communication network solutions for achieving efficient cooperation and coordination among players with different strategies in a networked N-player trust game represents an interesting research challenge.

- **Reward and Penalty Mechanism Design**: For the implementation of desirable social goals, the theory of *mechanism design* plays an important role. We have seen in the previous chapter that, according to [10], the theory of mechanism design can be thought of as the "engineering" side of economic theory. While the economic theorist wants to explain or forecast the social outcomes of mechanisms, the mechanism design theory reverses the direction of inquiry by identifying first the social goal and then asking whether or not an appropriate mechanism could be designed to attain that goal. And if the answer is yes, what form that mechanism might take, whereby a mechanism may be an institution, procedure, or game for determining desirable outcomes.

  An interesting example of mechanism design is the so-called *altruistic punishment* to ensure human cooperation in multiplayer public goods games [11]. Altruistic punishment means that individuals punish others, even though the punishment is costly and yields no material gain. It was experimentally shown that altruistic punishment of defectors (i.e. untrustworthy participants) is a key motive for cooperation in that cooperation flourishes if altruistic punishment is possible, and breaks down if it is ruled out. The design of externalities such as third-party punishment and alternative reward mechanisms for incentivizing human cooperation in multiplayer public goods games in general and N-player trust game in particular is of great importance.

- **Decentralized Implementation of Economic Experiments**: A widely used experimental software for developing and conducting almost any kind of economic experiments, including the aforementioned public goods games and our considered trust game, is the *Zurich Toolbox for Ready-made Economics (z-Tree)* [12]. The z-Tree software is implemented as a client-server application with a central server application for the experimenter, called z-Tree, and a remote client application for the game participants, called z-Leaf. It is available free of charge and allows economic experiments to be conducted via the Internet. At the downside, however, z-Tree does not support P2P communications between players, as opposed to a decentralized blockchain-enabled implementation.

7.3.2. Blockchain-Enabled Implementation

In this section, we develop a blockchain-enabled implementation of the classical trust game using Ethereum and experimentally investigate the beneficial impact of a simple yet effective blockchain mechanism known as *deposit* on enhancing both trust and trustworthiness as well as increasing social efficiency.

7.3.2.1. Experimenter Smart Contract

First, we develop a smart contract that replaces the experimenter in the middle between trustor and trustee (see Figure 7.1). The development process makes use of the Truffle framework, a decentralized application development framework. The resultant experimenter smart contract is written in the programming language Solidity. We then compile the experimenter smart contract into EVM byte code. Once the experimenter smart contract is compiled, it generates the EVM byte code and application binary interface (ABI). Next, we deploy the experimenter smart contract on Ethereum's official test network Ropsten. It can be invoked by using its address and ABI. More specifically, in our experimenter contract, we use the following global variables: (i) *msg.value*, which represents the transaction that is sent, and (ii) *msg.sender*, which represents the address of the player who has sent the transaction to the experimenter smart contract, i.e. trustor or trustee. Both trustor and trustee use their Ethereum externally owned account (EOA), which uses public and private keys to interact and invoke each function of our experimenter smart contract. In the following, we provide a brief overview of the core functions and parameters of our experimenter smart contract:

- **Function InvestFraction()**: This function allows the trustor to invest a portion p of her endowment X. Once called, it takes the received msg.value p from the trustor, multiplies it by factor K using the contract balance, and transfers it directly to the trustee's account. The trustee receives $msg.value \cdot K$.

- **Function SplitFraction()**: This function allows the trustee to split a portion q of the received investment from the trustor. Once called, it takes the set split amount from the trustee's account and sends it to the trustor's account.
- **Parameter Onlytrustor (modifier type)**: This modifier is applied to the *investFraction()* function. Thus, only the trustor can invoke this function of the experimenter smart contract.
- **Paramter Onlytrustee (modifier type)**: This modifier is applied to the *splitFraction()* function. Thus, only the trustee can invoke this function of the experimenter smart contract.

We note that after the execution of each function of the experimenter smart contract, an event is used to create notifications and saved logs. Events help trace and notify both players about the current state of the contract and activities.

7.3.2.2. Blockchain Mechanism Deposit

The use of one-way security deposits to provide trust for one party with respect to the other is quite common, particularly for the exchange of goods and services via e-commerce and crowdsourcing platforms. In the context of blockchains, a deposit is an agreement smart contract that defines the arrangement between parties, where one party deposits an asset with a third party. An interesting use case of the blockchain mechanism deposit was studied by Mut-Puigserver et al. [13]. The authors proposed a new protocol that achieves the fulfillment of all the desired properties of a registered e-Deliveries service using blockchain. In this protocol, the authors included a deposit mechanism with the aim to encourage the sender to avoid dishonest behavior and fraud attempts, and also to conclude the exchange in a predefined way following the phases of the protocol. The deposit will be returned to the sender if he finishes the exchange according to the protocol.

In our work, we propose to add an optional function *deposit()* to our experimenter smart contract to improve trust and trustworthiness between both players. Toward this end, we make the following two modifications:

- **Function Deposit()**: This function allows the trustee to submit an amount of $2 \leq D \leq X$ monetary units (i.e. Ether [ETH] in our considered case of Ethereum) as a deposit to the experimenter smart contract. The deposit is returned to the trustee only if a transaction with

$q > 0$ is completed. Otherwise, with $q = 0$, the trustee loses the deposit. It should be noted that the aforementioned *Onlytrustee()* modifier is also applied to this function.

- **Function SplitFraction()**: We make a modification to this function to allow the trustee to split the received amount (i.e. $q > 0$). Otherwise, the transaction is rejected until the trustee splits the received amount. Once this happens, the function transfers the amount to the trustor's account and returns the deposit D to the trustee's account.

7.3.2.3. Experimental Validation

Next, we investigate the impact of the deposit as an effective pre-commitment mechanism on the trust game performance via Ethereum-based blockchain experiments. We set $K = 2$ in our experimenter smart contract and consider different deposit values of $D = \{0, 2, 5, X\}$ ETH, whereby $D = 0$ denotes the classical trust game without any deposit. The experiment was conducted with two graduate students from different universities. The rationale behind the selection of only two students is to first focus on the conventional trust game that by definition involves only two players. This allows us to be more certain that the effects of the deposit mechanism are real. In addition, conducting our experiment with the same two participating students allows us to better observe the behavior change during the rounds of the game. As for our inclusion criteria, we note that the students didn't know each other's identity, which was important to ensure anonymity between them. Further, the students hadn't conducted any behavioral research experiment before. Nor did either participant have any prior knowledge or experience with the trust game or any other investment game experiments. The two participating students were male and their age was 23 and 25 years, respectively.

At the beginning of the experiment, both trustor and trustee were given an endowment of $X = 10$ ETH. We ran the experiment four times, each time for a different value of D. Each of the four experiments took five rounds. We note that for the experiment with $D = 10$ ETH, the trustee put her full endowment X into the deposit, thus $D = X$ ETH. All experiments were run across the Internet. Both participants interacted with our experimenter smart contract using their Ethereum accounts. We note that both the trustor and the trustee need to pay a gas fee. Gas refers to the pricing value, required to successfully conduct a transaction or execute a function in a smart contract on the Ethereum blockchain platform. Priced in small fractions of the cryptocurrency ETH, commonly referred to as Gwei. Each Gwei is equal to 0.000000001 ETH (i.e. 10^{-9} ETH). Given its lowest cost, we considered transaction fees associated with deploying the smart contract and sending transactions negligible compared to the amounts invested and split.

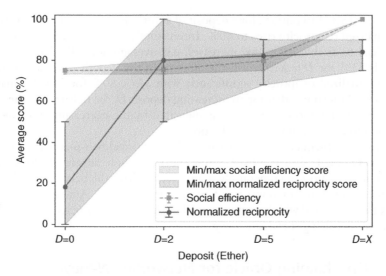

Figure 7.2 Average social efficiency and normalized reciprocity q/p vs. deposit $D = \{0,2,5,X\}$ Ether using experimenter smart contract with $K = 2$ and $X = 10$ (shown with minimum-to-maximum measured score intervals).
Source: Beniiche et al. (2021) © 2021 IEEE.

Figure 7.2 depicts the average social efficiency and normalized reciprocity (both given in percent) vs. deposit $D = \{0,2,5,X\}$ (given in ETH). We define normalized reciprocity as the ratio of q/p as a measure of the trustee's reciprocity, q, in response to the trustor's generosity, p. Note that the normalized reciprocity is useful to gauge the fair distribution of total payoffs from trustee to trustor, and vice versa, for a given achievable social efficiency. Note that Figure 7.2 also shows the interval between minimum and maximum measured score for each value of D.

We make the following interesting observations from Figure 7.2. First, the social efficiency continually grows for an increasing deposit D until it reaches the maximum of 100% for $D = X$. Thus, the social efficiency performance of the classical trust game can be maximized by applying the blockchain mechanism of deposit properly with $D = X$. This is due to the fact that the trustor sends her full endowment (i.e. $p = 1$) after the trustee has put in her maximum deposit. In doing so, a maximum total payoff of 30 ETH is achieved, translating into a social efficiency of 100%. It is worthwhile to mention that this was the case in all five rounds of the experiment. Second, the average normalized reciprocity improves significantly for increasing deposit D compared to the classical trust game without any deposit ($D = 0$). Specifically, in the classical trust game, the average normalized reciprocity is as low as 18%. By contrast, for a deposit of as little as $D = 2$ ETH, the average normalized reciprocity rises to 80%. Interestingly, further increasing D does not lead to sizable additional increases, e.g. average normalized reciprocity equals 83% for $D = X$. Hence, the amount of the deposit does not change the normalized reciprocity significantly with $q/p \approx 80\%$ for $D > 0$. Finally, Figure 7.2 illustrates that for an increasing deposit D, the behavior of the two players become more consistent, as indicated by the decreasing intervals of minimum to maximum measured scores.

In the subsequent section, we extend the classical two-player trust game to a networked N-player trust game and study how advanced blockchain technologies, most notably on-chaining oracles, drive the behavior of players by means of different reward and penalty mechanisms. Among others, we seek to understand whether an increased normalized reciprocity is achievable without sacrificing social efficiency.

7.4. On-Chaining Oracle for Networked *N*-Player Trust Game

Smart contracts need to acquire data about real-world states and events from outside the blockchain network, which cannot be achieved by smart contracts themselves because the blockchain environment is isolated from the external world. This is where oracles come into play to overcome the limitation by providing a link between off-chain and on-chain data. In general, oracles may get data from various sources. Depending on the respective source, oracles can be classified as follows:

- **Software Oracles**: Interact with online data sources and send information to the blockchain. The transmitted information may come from online databases, servers, websites, or, essentially, any data source on the web.
- **Hardware Oracles**: Designed to get information from the physical world and make it available to smart contracts. Such information may be relayed from sensors, Internet of things (IoT) devices, radio-frequency identification (RFID) tags, barcode/QR scanners, robots, or other information reading devices.
- **Human Oracles**: Individuals with specialized knowledge and skills in a particular field may also serve as oracles. They can research and verify the authenticity of information from various sources and transfer the information to smart contracts. Further, human oracles may provide smart contracts with answers to questions, report specified actions, or vote on the truth of a given event.

Oracles use a number of nodes to on-chain data onto a smart contract. These nodes define the trust model used by oracles [14]. Accordingly, oracles are classified into the following two models:

- **Centralized Oracles**: Controlled by a single source. The efficiency of centralized oracle is high, but their main problem is the existence of a single point of failure, where availability, accessibility, and level of certainty about the validity of the data depends on only one node.
- **Decentralized Oracles**: Resolve the singular point of failure problem of centralized oracles and thus increase the reliability of the information provided to smart contracts. A smart contract is able to query multiple oracles to determine the validity and accuracy of the data. This is why decentralized oracles are also referred to as consensus oracles.

Various oracle systems have been proposed, which aim at providing data to a blockchain reliably by incorporating techniques such as advanced cryptography, trusted execution environments, reporting, and voting [5].

7.4.1. Architecture of Oracle

Figure 7.3 depicts the architecture of our proposed on-chaining oracle for the networked N-player trust game. The proposed architecture comprises a set of clusters or pools. Each cluster contains three types of agent: (i) trustors, (ii) trustees, and (iii) observers. The difference between observers and players (trustors/trustees) is that observers don't play, but track and evaluate trust and trustworthiness criteria such as investment (p) and split (q). Players interact with the experimenter smart contract using their public–private keys through a decentralized application (DApp). The different rounds of the game are monitored remotely by the observers using *Etherscan.io*, an Ethereum blockchain explorer that uses the experimenter contract address and shows the different transactions between each pair of trustor and trustee in real time. We note that alternatively one may use *Alethio.io*, a monitoring tool that allows observers to send and receive alerts to and from any on-chain address, activity, or function.

The design of a third-party punishment and reward mechanism for incentivizing player cooperation in our networked N-player trust game is based on crowdsourcing. Specifically, observers provide their collective human intelligence to the nudge contract in order to punish a cluster or an individual player, who demonstrates inappropriate behavior, or provide a positive reward for good behavior. The nudge contract manages the reward–penalty mechanism in the form of loyalty points. A trustor can earn loyalty points for a honest transaction, investment, and engagement in the game and redeem earned points for rewards. Similarly, the trustee is rewarded for generous reciprocity.

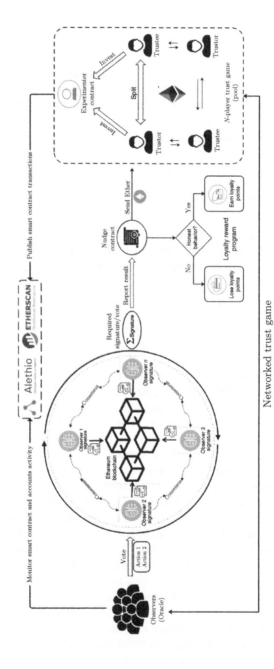

Figure 7.3   Architecture of on-chaining oracle for networked *N*-player trust game.
Source: Beniiche et al. (2021) © 2021 IEEE.

Loyalty points keep the players engaged and aware of the overall goals, i.e. increase of total payoff, social efficiency, and normalized reciprocity. In addition, the players have a score profile associated with their public key, whereby players earn 1 point for every honest action and lose 1 point if their action is dishonest. The scoring profile is managed by the nudge contract. Trustor and trustee can check the status of their loyalty reward points by calling the function *getTrustorLoyalty*() and *getTrusteeLoyalty*(), respectively. Furthermore, an incentive strategy was designed to incorporate principles of behavioral psychology using economic outcomes to render the system more effective in changing the players' behavior. Players earn a monetary reward in the form of ETH after reaching a certain number of loyalty reward points in the game, e.g. 10 points = 1 ETH. The ETHs earned are added to the player's endowment X, which will be used for the investment and payoff in future rounds of the game.

7.4.2. On-Chaining of Voting-Based Decisions

In our oracle implementation, we assigned predetermined public keys to both players and observers. The creation of each key pair can be accomplished by using several options, including Ethereum wallets and online/offline Ethereum address generators, e.g. Vanity-ETH. All public keys are declared in the nudge contract, whose purpose is to allow only registered observers to vote while automatically rejecting malicious voters. To facilitate the formation of a majority, the number of possible voting options is restricted to the four following functions: *VOTE_RewardTrustor*, *VOTE_RewardTrustee*, *VOTE_PunishTrustor*, and *VOTE_PunishTrustee*. Further, to ensure a trustworthy on-chaining decision, a k-out-of-M threshold signature is used to reach a consensus on the function to be executed. A k-out-of-M threshold signature scheme is a protocol that allows any subset of k players out of M players to generate a signature, and disallows the creation of a valid signature if fewer than k players should participate. The right decision is determined as the one that has received the desired number of votes. Once the function is executed, the nudge contract allocates the reward or punishment loyalty points to each player who behaved in a trusted or untrusted way, respectively.

7.4.3. Experimental Validation

We compare the performance of our proposed on-chaining oracle for the multiplayer N-player trust game with the conventional two-player baseline experiment. Toward this end, we invited the same two students, who have played the classical two-player trust game before, and asked them to play the game again, i.e. without any observers. Next, we invited them to play the game

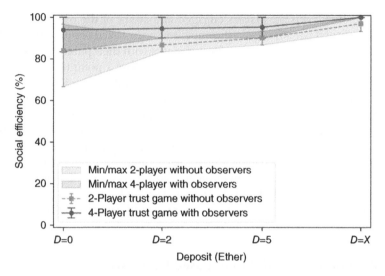

Figure 7.4 Average social efficiency vs. deposit $D = \{0, 2, 5, X\}$ ETH for 2-player trust game without observers and 4-player trust game with observers (shown with minimum-to-maximum measured score intervals).
Source: Beniiche et al. (2021) © 2021 IEEE.

in the presence of two observers. The two players were informed that their account is associated with loyalty reward points, which will be increased if they act honestly. Otherwise, they will be punished and lose 1 loyalty point. Both players were aware that they will be rewarded with 1 ETH for each 10 accumulated loyalty reward points. In addition, they are notified that the decision will be made by two observers, who will monitor their online transactions in order to make their independent reward/penalty decisions. All four participants interact anonymously via the Internet.

Figure 7.4 compares the average social efficiency of the two-player trust game without observers with that of the four-player trust game with observers. The figure clearly demonstrates the beneficial impact of the presence of observers on social efficiency for all values of D. Note that with observers the instantaneous social efficiency reaches the maximum of 100% for all values of D, as opposed to the two-player trust game where this occurs requiring the full deposit of $D = X$ ETH. As for the normalized reciprocity achievable with and without observers things are similar, as shown in Figure 7.5. However, while the presence of observers helps raise the average (and instantaneous) normalized reciprocity consistently above 80% (compared to below 80% in Figure 7.2), there still remains room for further improvement, especially for $0 \leq D < X$.

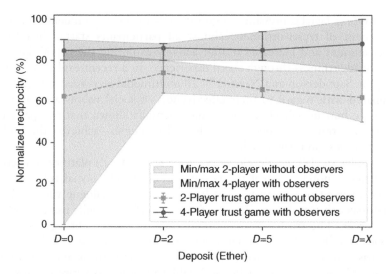

Figure 7.5 Average normalized reciprocity q/p vs. deposit $D = \{0, 2, 5, X\}$ ETH for 2-player trust game without observers and 4-player trust game with observers (shown with minimum-to-maximum measured score intervals).
Source: Beniiche et al. (2021) © 2021 IEEE.

7.5. Playing the *N*-Player Trust Game with Persuasive Robots

Many studies have shown that the physical presence of robots benefits a variety of social interaction elements such as persuasion, likeability, and trustworthiness. Thus, leveraging these beneficial characteristics of social robots represents a promising solution toward enhancing the performance of the trust game. Social robots connected with human operators form a physical embodiment that creates the new paradigm of an immersive coexistence between humans and robots, whereby persuasive robots aim at changing the behavior of users through social influence. Importantly, these robots are less like tools and more like partners, whose persuasive role in a social environment is mainly human-centric [15].

Recently, in [16], an experimental pilot study with five participants adapted the trust game from its original human–human context to an sHRI setting using a humanoid robot operated in a *Wizard-of-Oz* (WoZ) manner, where a person controls the robot remotely. The obtained findings suggest that people playing the social human–robot interaction (sHRI) trust game follow a human–robot trust model that is quite similar to the human–human trust model. However, due to the lack of common *social cues* present in humans (e.g. facial expressions or gestures) that generally influence the initial assessment of trustworthiness, almost all participants started investing a lower amount and increased it after actively exploring the robot's behavior and trustworthiness through social experience. In a recent exploratory sHRI study by Saunderson and Nejat [17], 10 multimodal *persuasive strategies* were compared with regard to their effectiveness of social robots attempting to influence human behavior. It was experimentally shown that two particular persuasive strategies – affective and logical strategies – achieved the highest persuasiveness and trustworthiness.

Similar to [18], we developed a *Crowd-of-Oz* (CoZ) platform for letting observers remotely control the gestures of Softbank's social robot Pepper placed in front of the trustee and have a real-time dialogue via web-based text-to-speech translation. The CoZ user interface is built using a Django web server. The trustee can communicate with Pepper through voice and Pepper's tactile tablet. To support voice communication, we implemented a web-based speech-to-text tool. When text is extracted from voice, the trustee can see his/her message on Pepper's tablet in order to verify it. Next, the speech-to-text function calls another function to add additional fields to the main message (extracted text), including sequence ID, sender ID, message type, and time to make the message distinguishable on the Django server. The called function executes a marshaling process and sends the message to the Django server through the OOCSI middleware. The OOSCI middleware is a message-based connectivity layer and is platform-independent inspired by the concept of remote procedure call (RPC) for connecting web clients.

In our developed CoZ system, there are two types of message: information and control. The information messages are created by the observers. This type of message is multicast to all observers and the trustee through Pepper to update them, but not the trustor. The trustee can see all the information messages on Pepper's tablet. Moreover, Pepper uses a text-to-speech function to transfer the observers' messages to the trustee. The control messages are used for important functionalities of the CoZ architecture, e.g. performing a gesture on Pepper. When an observer presses a social cue button, the CoZ web-interface invokes a JavaScript method to call a new event on the Django server. The JavaScript method creates a control message, performs the marshaling process, and sends it to the Django server using OOCSI. On the Django server, the first step is unmarshaling the message. Next, it invokes the related method to perform selected social cues on Pepper. The invoked method sets all the related joints' angles plus the light-emitting diode (LED) colors of Pepper's eyes. Given that two or more observers may press the same or different social cue buttons simultaneously, the Django server implemented a queue to synchronize all issued commands. While Pepper is performing a gesture, the Django server puts the next gesture in the queue and sends it to Pepper back-to-back.

Further, our CoZ user interface provides a section, where an observer can watch the trustee's environment through Pepper's eyes. To implement this part, we used OpenCV, Flask, and CV2, which are all Python libraries. The Django server invokes a method on Pepper called "ALVideoDevice" to start recording videos. By using CV2 we convert the video frames to a matrix of 320-width and 240-height (desired image size), whereby the value of each element is in the range of 0 to 255 (indicating the color of each pixel's image, also known as red green blue [RGB]). Subsequently, the matrix is converted to a PNG image. This process runs in an endless loop and only terminates if triggered by the Django server (e.g. end of experiment). Next, the Flask server stores the sequence of produced images with a valid URL. To make live video streams accessible over the Internet we used virtual private network (VPN). Moreover, in our CoZ interface, we used an IFrame (Inline Frame) tag to demonstrate live video streaming using the valid URL. An IFrame is an HTML document embedded inside another HTML document on a website. The IFrame HTML element is often used to insert content from another source, such as a camera, into a Web page. In our CoZ user interface, we also realized four buttons to turn Pepper's head to left, right, up, and down. When an observer presses one of these buttons, the CoZ interface runs a JavaScript to invoke a method to create a control message, marshaling process, and send it to the Django server. Upon reception, the Django server performs unmarshaling to extract the main message and then invokes the "ALMotion" along with initializing some parameters like speed, angle, and joint name. For each invocation, Pepper turns her head by 10°.

The user CoZ interface also displays nine social cue buttons to prevent possible typos and save time for observers to fill communication gaps. The nine social cue buttons were as follows: "Gain time," "Tell me about it," "Good job," "Hi," "Bye," "Open arms," "Taunting hands," "No," and "Ask for attention." Observers may press to perform different gestures of Pepper during conversation and thereby influence the trustee's behavior. In addition, we drafted two scripts, one for a logical persuasive strategy appealing to the left side of the brain (i.e. logics) and another one for an affective persuasive strategy appealing to the right side of the brain (i.e. emotions) of the trustee. Each script contains prespecified sentences stored in pull-down menus in the CoZ interface, from which observers may choose in order to nudge the trustee's behavior toward reciprocity via real-time text-to-speech messages. The different persuasive robot strategies operate as follows:

- **Logical Strategy**: Contains a set of reward/punishment mechanisms. In addition, Pepper performs some economical and technical advice via text-to-speech through the above described CoZ platform.
- **Affective Strategy**: Contains a set of reward/punishment mechanisms and Pepper uses text-to-speech encouragement messages through the

CoZ platform. In addition, Pepper shows social cues by means of gestures and embodied communication toward the trustee.

■ **Mixed Strategy**: Combines the above logical and affective strategies into one mixed strategy. It contains a set of reward/punishment mechanisms and Pepper provides not only economical and technical advice but also encouragement via text-to-speech messages through the CoZ platform. In addition, Pepper shows social cues by means of gestures and embodied communications toward the trustee.

For illustration, Table 7.1 lists the social cues used by Pepper in our proposed mixed logical–affective persuasive strategy. In this strategy, one observer plays the logical strategy and the other observer plays the affective strategy such that the trustee receives mixed messages and mixed embodied communications. Depending on the trustee's behavior, the observers carries out the "Trusted behavior action" or the "Untrusted behavior action" in each round of the experiment. The social cues in Table 7.1 enable the observers to control Pepper's text-to-speech and embodied communications using our developed CoZ platform.

We ran large-scale experiments involving 20 students to measure the effectiveness of our developed persuasive robotics strategies. Similar to our last experiment in the two players' trust game, the participating students didn't know each other's identity. Also, students hadn't conducted any behavioral research experiment before. The age of the selected students was between 24 and 32 years. Three students were female and seventeen students were male. The experiment was divided into four trials: baseline, logical, affective, and mixed strategy. Each trial involved five rounds. We first conducted a baseline trust-game experiment, where trustees didn't interact with Pepper, as done previously, followed by experiments exposing trustees to Pepper's logical, affective, and mixed logical-affective persuasive strategies. Both trustor and trustee interacted via a blockchain account with the experimenter's smart contract. The trustor played the game from a separate room, while the trustee was in the lab alone with Pepper. Pepper was controlled via our CoZ platform remotely by the observer. We used the same parameter settings, i.e. endowment $X = 10$ ETH for the trustor and $K = 2$. Further, in all persuasive strategies, we didn't use any deposit mechanism (i.e. $D = 0$).

Figure 7.6 demonstrates the superior effectiveness of our persuasive strategies, especially mixed ones appealing to both sides of the brain, resulting in average normalized reciprocity well above 100%. Further, to better reveal the differences among the persuasive strategies, we have calculated the measurement range for the four strategies. The measurement range for the

Table 7.1   Social cues used by Pepper in mixed persuasive strategy.

Round number	Trusted behavior action	Untrusted behavior action
Round 1	*Text-to-speech*: Trust game is a cooperative investment game. You all play together to get the best total payoff!	Untrusted behavior will be shown in Round 2
Round 2	*Text-to-speech*: Awesome! That's a split worth celebrating!	*Text-to-speech*: If this behavior is repeated, you will receive a punishment from the observers. *Embodied communication*: Taunting hand gesture
Round 3	*Text-to-speech*: If this good behavior is repeated, your partner will invest more in the next round. *Embodied communication*: Open arm gesture	*Text-to-speech*: Weak reciprocity can cause costly punishment for you. *Embodied communication*: Taunting hand gesture
Round 4	*Text-to-speech*: Incredible! Your partner must be impressed! *Embodied communication*: Open arm gesture	*Text-to-speech*: With such behavior, the punishment will be executed next round. *Embodied communication*: Taunting hand gesture
Round 5	*Text-to-speech*: Congrats! Your good behavior toward your partner has provided you with an incremental total payoff over all rounds of the game. *Embodied communication*: Open arm gesture	*Text-to-speech*: Your bad behavior translated into a very weak total payoff. *Embodied communication*: Taunting hand gesture

Source: Beniiche et al. (2021) © 2021 IEEE.

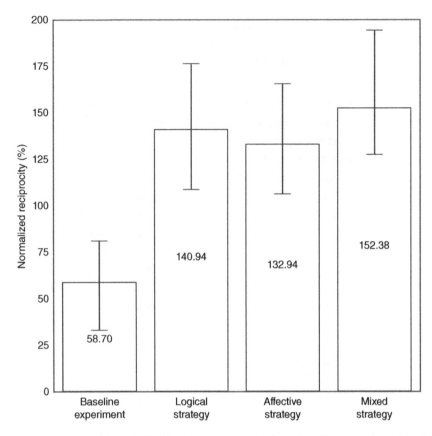

Figure 7.6 Average normalized reciprocity q/p without (baseline experiment) and with using logical, affective, and mixed logical-affective persuasive strategies for $D = 0$ (shown with minimum-to-maximum measured score intervals). © 2021 IEEE. *Source: Beniiche et al. (2021)*

baseline experiment is 48.2 (max = 81, min = 32.8), while for the logical strategy it is 67.8 (max = 176.4, min = 108.6), for the affective strategy it is 59.4 (max = 165.6, min = 106.2), and for the mixed strategy it is 67 (max = 194.4, min = 127.4). As the results show, the baseline experiment has the smallest measurement range. Next, we computed the standard deviation for the baseline experiment as well as logical, affective, and mixed strategies, which is equal to 15.6, 21.75, 21.10, and 22.73, respectively. The results show that the baseline experiment has the smallest standard deviation among all considered strategies, while the mixed strategy has the largest one. Finally, we have computed the variance for the persuasive strategies under consideration.

The calculated variance equals 245.83, 473.17, 445.25, and 517.03 for the baseline, logical, affective, and mixed strategy, respectively. Based on the gathered results, we observe that the baseline experiment has the smallest and the mixed strategy has the largest variance.

7.6. Tokenomics

Cardenas and Kim [1] provided insights into how blockchain and other decentralized technologies have an impact on the interaction of humans with robot agents and their social integration into human society. Toward this end, blockchain technologies can serve as a ledger, where robots and humans may access and record anything of value, such as ownership titles or financial transactions. Further, smart contracts help encode the self-enforceable and self-verifiable agreement logic between a robot and a human. Cryptocurrencies – not only coins but also tokens – may be used to allow robots to hold financial obligations and enter into exchanges of value with a human, and vice versa.

In this chapter, we investigated advanced blockchain technologies such as oracles that enable the on-chaining of blockchain-external off-chain information stemming from human users. In doing so, oracles leverage on human intelligence rather than machine learning only. Our empirical results demonstrated that the presence of third parties such as human observers and in particular social robots play an important role in the blockchain-enabled trust game, a cornerstone of behavioral economics. The peer pressure executed by on-chaining oracles and especially the embodied communication enabled by persuasive robots were shown to have a potentially greater social impact than monetary incentives such as the considered cryptocurrency ETH.

Recall from Chapter 1 that the Web3 token economy is the successor of the Web1 knowledge economy and today's Web2 platform economy (see also Figure 1.2). The token economy enables completely new use cases, business models, and types of assets and access rights in a digital way that were economically not feasible before, thus enabling completely new use cases and value creation models. In a token economy, anyone's contribution is compensated with a token. Unlike coins, which have been typically used only as a payment medium, tokens may serve a wide range of different non-monetary purposes as well as incentives for an autonomous group of individuals to collaborate and contribute to a common goal.

The shift from conventional monetary economics to non-monetary tokenomics and the central role tokens play in blockchain-based ecosystems were analyzed recently by Freni et al. [19]. Since the ground-breaking innovation of the Bitcoin and its underlying "chain of blocks" (later renamed as blockchain) introduced by Satoshi Nakamoto[1] in a white paper published on 31 October 2008, disintermediation and decentralization have been applied way beyond the mere monetary context, embracing different value-based scenarios, through tokenization. Once tokenized, every kind of value can be managed as a digital asset, whose unit of account is a dedicated token. The tokens can be minted by any individual or organization that defines the set of rules governing them, such as the token features, monetary policy, and users' incentive system. As a result, tokenization can be described as the creation of a self-governed (tok)economic system, whose rules are programmed by the token issuer.

In conventional economics, innovation proceeds and propagates by introducing a change in the context of set rules and by observing how such a relatively rigid framework reacts to the change. Therefore, the outcome of the introduced innovation is assessed, at first, on a predictive basis. Conversely, in tokenomics, innovation is put forward by designing the rules governing the playground in a way that the stakeholders' behavior aligns with the goal pursued. Hence, the resultant paradigmatic shift from economics to tokenomics moves from the passive observation of the ecosystem's reaction to a change toward the active design of the ecosystem's constituent laws that are aimed at reaching the desired outcome [19].

References

1. I. S. Cardenas and J.-H. Kim. Robonomics: The Study of Robot-Human Peer-to-Peer Financial Transactions and Agreements. In *Proc., HRI' 20: Companion of the 2020 ACM/IEEE International Conference on Human-Robot Interaction*, pages 8–15, March 2020.
2. A. Beniiche, S. Rostami, and M. Maier. Robonomics in the 6G era: playing the trust game with on-chaining oracles and persuasive robots. *IEEE Access*, 9:46949–46959, March 2021.
3. O. Novo. Blockchain meets IoT: an architecture for scalable access management in IoT. *IEEE Internet of Things Journal*, 5(2):1184–1195, April 2018.
4. J. Adler, R. Berryhill, A. Veneris, Z. Poulos, N. Veira, and A. Kastania. ASTRAEA: A Decentralized Blockchain Oracle. In *Proc., IEEE International Conference on Cyber, Physical and Social Computing*, pages 1145–1152, July/August 2018.
5. J. Heiss, J. Eberhardt, and S. Tai. From Oracles to Trustworthy Data On-chaining Systems. In *Proc., IEEE International Conference on Blockchain*, pages 496–503, July 2019.

---

[1] Satoshi Nakamoto is the name used by the presumed pseudonymous person or persons who wrote the original Bitcoin white paper titled "Bitcoin: A Peer-to-Peer Electronic Cash System."

6. J. Berg, J. Dickhaut, and K. McGabe. Trust, reciprocity, and social history. *Games and Economic Behavior*, 10(1):122–142, July 1995.

7. C. Alós-Ferrer and F. Farolfi. Trust games and beyond. *Frontiers in Neuroscience*, 13(Article 887):1–14, September 2019.

8. H. Abbass, G. Greenwood, and E. Petrak. The N-player trust game and its replicator dynamics. *IEEE Transactions on Evolutionary Computation*, 20(3):470–474, June 2016.

9. M. Chica, R. Chiong, M. Kirley, and H. Ishibuchi. A networked N-player trust game and its evolutionary dynamics. *IEEE Transactions on Evolutionary Computation*, 22(6):866–878, December 2018.

10. S. Maskin. Mechanism design: how to implement social goals. *The American Economic Review*, 98(3):567–576, June 2008.

11. E. Fehr and S. Gächter. Altruistic punishment in humans. *Nature*, 415:137–140, January 2002.

12. U. Fischbacher. z-Tree: Zurich toolbox for ready-made economic experiments. *Experimental Economics*, 10(2):171–178, June 2007.

13. M. Mut-Puigserver, M. A. Cabot-Nadal, and M. M. Payeras-Capella. Removing the trusted third party in a confidential multiparty registered eDelivery protocol using blockchain. *IEEE Access*, 8:106855–106871, June 2020.

14. H. Al-Breiki, M. H. U. Rehman, K. Salah, and D. Svetinovic. Trustworthy blockchain oracles: review, comparison, and open research challenges. *IEEE Access*, 8:85675–85685, May 2020.

15. M. Siegel, C. Breazeal, and M. I. Norton. Persuasive Robotics: The Influence of Robot Gender on Human Behavior. In *Proc., IEEE/RSJ International Conference on Intelligent Robots and Systems*, pages 2563–2568, October 2009.

16. R. C. R. Mota, D. J. Rea, A. L. Tran, J. E. Young, E. Sharlin, and M. C. Sousa. Playing the 'Trust Game' with Robots: Social Strategies and Experiences. In *Proc., IEEE International Symposium on Robot and Human Interactive Communication*, pages 519–524, August 2016.

17. S. Saunderson and G. Nejat. It would make me happy if you used my guess: comparing robot persuasive strategies in social human-robot interaction. *IEEE Robotics and Automation Letters*, 4(2):1707–1714, April 2019.

18. T. Abbas, V.-J. Khan, and P. Markopoulos. CoZ: A crowd-powered system for social robotics. *Elsevier SoftwareX*, 11(100421):1–7, January-June 2020.

19. P. Freni, E. Ferro, and R. Moncada. Tokenization and Blockchain Tokens Classification: A Morphological Framework. In *Proc., IEEE Symposium on Computers and Communications (ISCC)*, pages 1–6, July 2020.

CHAPTER 8

Society 5.0: Internet As If People Mattered

"If people did not sometimes do silly things, nothing intelligent would ever get done."

LUDWID WITTGENSTEIN
(1889–1951)

While the primary focus of 5G has been on industry verticals, future 6G mobile networks are anticipated to become more human-centered. Toward this end, it is important to take a number of different factors into account in order to develop a more realistic understanding of human nature that challenges that of rational individuals driven by self-interest, as traditionally assumed in mainstream economics. Emerging cyber-physical-social system (CPSS) aim at functionally integrating human beings into today's cyber–physical systems (CPS) at the social, cognitive, and physical levels. CPSS are instrumental in realizing the human-centered Society 5.0 vision. Society 5.0 envisions human beings to increasingly interact with social robots and embodied artificial intelligence (AI) in their daily lives. In this chapter, we expand on our work on robonomics and tokenomics discussed in Chapter 7. After introducing our CPSS-based bottom-up multilayer token engineering framework for Society 5.0, we experimentally demonstrate how the collective human intelligence of a blockchain enabled decentralized autonomous organization (DAO) can be enhanced via purpose-driven tokens [1].

6G and Onward to Next G: The Road to the Multiverse, First Edition. Martin Maier.
© 2023 The Institute of Electrical and Electronics Engineers, Inc.
Published 2023 by John Wiley & Sons, Inc.

8.1. Human Nature: Bounded Rationality and Predictable Irrationality

Recall from Chapter 6 that behavioral economics assumes that the behavior of individuals is impacted by a number of different factors than economic rationality, such as psychological, emotional, cultural, cognitive, and social factors. In fact, people make over 90% of their decisions based on mental shortcuts or rules of thumb. Especially, under pressure and in situations of high uncertainty, humans tend to rely on anecdotal evidence and stereotypes to help them understand and respond to events more quickly. This is due to the fact that the rationality of individuals and institutions is bounded by time and cognitive limitations and that good enough solutions are preferred over perfect solutions. *Bounded rationality* is a concept proposed by Herbert Simon – recipient of the Nobel Memorial Prize in Economic Sciences in 1978 and the Turing Award in computer science in 1975 – that challenges the notion of human rationality as implied by the concept of *homo oeconomicus*. Rationality is bounded because there are limits to our thinking capacity, available information, and time.

Behavioral economics builds on the learnings of cognitive psychology, a field of psychology that studies mental processes. Alternative economic theories, such as behavioral economics, are based on the assumption that individual action is more complex. Behavioral economics studies why market actors are economically irrational and how others can profit from such *predictable irrationality*. It is important to note that these irrational behaviors are neither random nor senseless. They are systematic and predictable, making us predictably irrational [2].

The relation between the Internet and other technosocial systems that a society uses and human behavior can be either simple (i.e. linear cause-and-effect relationships) or complex (i.e. nonlinear and hard-to-predict relationships). We have seen in Chapter 6 that token systems, if designed properly, are able to steer collective action within the network. The definition of a clear purpose of the token is necessary for the design process. The clearer the purpose, the more resilient the network. Generally speaking, a token should only have one purpose. If you have multiple purposes, you probably need more token types. Otherwise, the mechanism design of your token system can become too complex. Once the purpose is defined, one can derive the properties of the token, e.g. fungibility, transferability, or expiry date (see also Table 6.1). The properties of a token are the basis for modeling a fault-tolerant mechanism to steer the network toward a collective goal. The aim of such a fault-tolerant mechanism is to define upon which behavior the tokens are minted, such that it is resilient against corruption, attacks, or mistakes. Proof of work (PoW) has proven to be resilient to achieve the purpose.

Importantly, if we fail to incorporate ethical questions in the design thinking process of such token systems, we will create protocol biases. If this is done after a token system has been created, these biases are hard to reverse due to system inertia. However, we do not have to reinvent the wheel. We can apply *engineering ethics* to the creation of Internet-based systems, something that the Silicon Valley and other big players of the Internet era have failed to do [3].

8.2. Society 5.0: Human-Centeredness

In Chapter 1, we have discussed how the current fourth industrial revolution has been enabled through the Internet of things (IoT) in association with other emerging technologies, most notably CPS. CPS help bridge the gap between manufacturing and information technologies and give birth to the smart factory. This technological evolution ushers in the Industry 4.0 as a prime agenda of the High-Tech Strategy 2020 Action Plan taken by the government of Germany, the Industrial Internet from General Electric in the United States, and the Internet+ from China. One of the most important paradigmatic transitions characterizing Industry 5.0 is the shift of focus from technology-driven progress to a thoroughly human-centric approach. An important prerequisite for Industry 5.0 is that technology serves people, rather than the other way around, by expanding the capabilities of workers (up-skilling and re-skilling) with innovative technological means such as virtual reality (VR)/augmented reality (AR) tools, mobile robots, and exoskeletons. Note that while Industry 4.0 implies minimum human intervention due to automation, Industry 5.0 prioritizes augmentation of the human over automation of the human.

For illustration, Figure 1.4 in Chapter 1 depicted the transition from past to future societies and their coevolution with industry. While the focus of Society 4.0 was on building an information society via information and communications technologies (ICT) for the purpose of increasing profitability, Society 5.0's main goal is to merge cyberspace and physical space for the purpose of advancing humanity. Society 5.0 seeks to revolutionize not only the industry through ICT but also the living spaces and habits of the public. Society 5.0 focuses heavily on the public impact of technology and aspires to create a supersmart society, thus requiring metrics that are much more complex than those used in Industry 4.0, whose focus centers on minimizing manufacturing costs without taking social issues into account [4].

According to [5], Society 5.0 seeks to create a sustainable society for human security and well-being, which is aligned with the United Nations' sustainable development goals (SDGs), and to create new values by making connections between people and things and between the real and cyber worlds through digitalization. Shiroishi et al. [5] emphasize that digitalization is only the means and that it is essential that we humans remain the central actors so that a firm focus is kept on building a society that makes us happy. More specifically, Matsuda et al. [6] provide the following definition of Society 5.0: "A human-centered society that balances economic advancement with the resolution of social problems by a system that highly integrates cyberspace and physical space, a system which consists of physically controllable technologies such as AI, robots, and IoT integrated into cyberspace." To better understand the differences between Industry 4.0/5.0 and Society 5.0, Chaudhari et al. [7] compare them in terms of involved technologies, research areas, power source, motto as well as their underlying motivation. They argue that while the main motivation behind Industry 4.0 and Industry 5.0 is mass production and a smart sustainable society, respectively, Society 5.0's main motivation is that humankind live in harmony with nature.

In the following, we aim at driving the *bionic convergence* of robonomics, DAO, and the Internet of no things as our CPSS of choice to advance the collective intelligence (CI) of Society 5.0 in the next-generation Internet known as Web3. Given that token design is still in the very early research stages as explained in Chapter 6, we pay particular attention to the important problem of token engineering and present our CPSS-based bottom-up token engineering DAO framework for Society 5.0, including its multilayer architecture and mechanism design. Importantly, we experimentally demonstrate the potential of the biological *stigmergy* mechanism for advancing CI in a CPSS-based DAO via tokenized digital twins.

8.3. The Path (DAO) to a Human-Centered Society

8.3.1. Token Engineering DAO Framework for Society 5.0

Recently, Wang et al. [8] proposed a multilayer DAO reference model for the token design, though it was intentionally kept generic without any specific relation to Society 5.0. The bottom-up architecture of the DAO reference model comprises the following five layers: (i) basic technology, (ii) governance operation, (iii) incentive mechanism, (iv) organization form, and (v) manifestation. We refer the interested reader to [8] for further information on the generic DAO reference model and a more detailed description of each layer. In the following, we adapt the generic DAO reference model to the specific requirements of Society 5.0 and highlight the modifications made in our CPSS-based bottom-up token engineering DAO framework.

Figure 8.1 depicts our proposed multilayer token engineering DAO framework for Society 5.0 that builds on top of state-of-the-art CPSS. While the IoT as prime CPS example has ushered in Industry 4.0, advanced CPSS

Token engineering DAO framework for Society 5.0

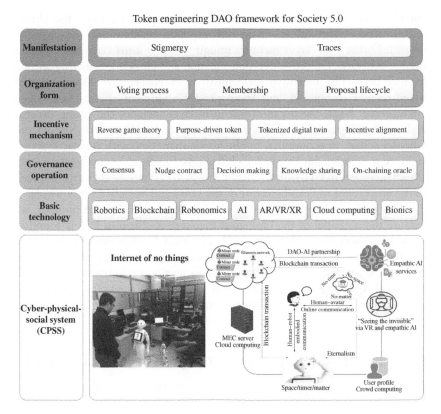

Figure 8.1   CPSS-based bottom-up token engineering DAO framework for Society 5.0.
Source: Beniiche et al. (2022) © 2022 IEEE.

such as the future Internet of no things, briefly mentioned in Chapter 1, will be instrumental in ushering in Society 5.0. Recall that the Internet of no things creates a converged service platform for the fusion of digital and real worlds that offer all kinds of human-intended services without owning or carrying any type of computing or storage devices. It envisions Internet services to appear from the *surrounding environment* when needed and disappear when not needed.

The basic technology layer at the bottom of Figure 8.1 illustrates the key enabling technologies (e.g. blockchain) underlying the Internet of No Things. In addition, this layer contains future technologies, most notably bionics, that are anticipated to play an increasingly important role in a future super smart Society 5.0 (see also Figure 1.4). Above the basic technology layer, there exists the governance operation layer. Generally speaking, this layer encodes consensus via smart contracts (e.g. voting) and realizes the DAO's self-governance through on-chain and off-chain collaboration (on-chaining oracle). Further, this layer includes nudging mechanisms via smart contract (nudge contract), collective decision-making, and knowledge-sharing among its members. The incentive mechanism layer covers the aforementioned token-related techniques and their proper alignment to facilitate token engineering. Next, the organization form layer includes the voting process and membership during the lifecycle of a proposed DAO project. Recall from Chapter 6 that, in economics, public goods that come with regulated access rights (e.g. membership) are called club goods. Finally, the manifestation layer allows members to take simple, locally independent actions that together lead to the emergence of CAS behavior of the DAO and Society 5.0 as a whole.

Due to its striking similarity to decentralized blockchain technology, we explore the potential of the biological *stigmergy* mechanism widely found in social insect societies such as ants and bees, especially their inherent capability of self-organization and indirect coordination by means of olfactory traces that members create in the environment. Upon sensing these traces, other society members are stimulated to perform succeeding actions, thus reinforcing the traces in a self-sustaining autocatalytic way without requiring any central control entity, as explained in Section 8.4.

8.3.2. The Human Use of Human Beings: Cybernetics and Society

A prominent example of moving a Web2-based social network to Web3 was Facebook's recent announcement in June 2019 to launch a new infrastructure to manage their own token coined *Libra* (later renamed Diem), including suitable price stability mechanisms for its exchange with fiat currencies. Facebook's recently launched Metaverse vision will support non-fungible tokens (NFTs) to build a creator economy, where content creators can make a living from online contributions. It is a good example of the emerging token economy. However, the design of tokenized currencies will not be sufficient for realizing the Society 5.0 vision. To see this, note that people, including former and founding executives, began publicly questioning the impact of social media on our lives and opened up about their regrets over helping create social media as we know it today.[1] For instance, during a public discussion at the Stanford Graduate School of Business, Chamath Palihapitiya, former vice president of user growth at Facebook, told the Stanford audience that the tools we have created are ripping apart the social fabric of how society works. It is eroding the core foundation of how people behave by and between each other. He concluded that he does not have

---

[1] https://gizmodo.com/former-facebook-exec-you-don-t-realize-it-but-you-are-1821181133

a good solution. His solution is just that he does not use these tools anymore, nor are his kids allowed to do so. Or as Facebook's first president, Sean Parker famously put it: "God only knows what it's doing to our children's brains."

Useful hints for more human-centered solutions may be found in the origins of cybernetics. In his seminal book "The Human Use of Human Beings: Cybernetics and Society," Norbert Wiener, the founder of cybernetics, argues that the danger of machines working on cybernetic principles, though helpless by themselves, is that such machines may be used by a human being or a block of human beings to increase their control over the rest of the human race. In order to avoid the manifold dangers of this, Wiener emphasizes the need for the anthropologist[2] and the philosopher. He postulates that scientists must know what man's nature is and what his built-in purposes are, arguing that the integrity of *internal communication* via feedback loops is essential to the welfare of society.

8.4. From Biological Superorganism to Stigmergic Society

In this section, we explore how the aforementioned integrity of internal communication may be achieved in Society 5.0 by borrowing ideas from the biological superorganism with brain-like cognitive abilities observed in colonies of social insects. The concept of stigmergy (from the Greek words *stigma* "sign" and *ergon* "work"), originally introduced in 1959 by French zoologist Pierre-Paul Grassé, is a class of self-organization mechanisms that made it possible to provide an elegant explanation to his paradoxical observations that in a social insect colony individuals work as if they were alone while their collective activities appear to be coordinated. In stigmergy, traces are left by individuals in their environment that may feedback on them and thus incite their subsequent actions. The colony records its activity in the environment using various forms of storage and uses this record to organize and constrain collective behavior through a feedback loop, thereby giving rise to the concept of *indirect communication*. As a result, stigmergy maintains social cohesion by the coupling of environmental and social organization. Note that with respect to the evolution of social life, the route from solitary to social life might not be as complex as one may think. In fact, in the AI subfield of swarm intelligence, e.g. swarm robotics, stigmergy is widely recognized as one of the key concepts.

---

[2] See also [9] on "Who Will be the Members of Society 5.0? Towards an Anthropology of Technologically Posthumanized Future Societies."

8.4.1. Stigmergy Enhanced Society 5.0: Toward the Future Stigmergic Society

In the following, we illustrate how our CPSS-based token engineering DAO framework in Figure 8.1 can be applied to Society 5.0 and describe the involved bottom-up design steps of suitable purpose-driven tokens and mechanisms:

- **Step1: Specify Purpose**
  Recall from Chapter 6 that the design of any tokenized ecosystem starts with a desirable output, i.e. its purpose. As discussed in Chapter 1, the goal of Society 5.0 is to provide the technosocial environment for CPSS members that (i) extends human capabilities and (ii) measures activities toward human co-becoming super smart. Toward this end, we advance AI to CI among swarms of connected human beings and things, as widely anticipated in the 6G and Next G era.

- **Step 2: Select CPSS of Choice**
  We choose our recently proposed Internet of no things as state-of-the-art CPSS, since its final transition phase involves nearables that help create intelligent environments for providing human-centered and user-intended services. Recall from Chapter 5 that our extrasensory perception network (ESPN) integrates ubiquitous and persuasive computing in nearables (e.g. social robot, virtual avatar) to change the behavior of human users through social influence. Here, we focus on blockchain and robonomics as the two basic technologies to expand ESPN's online environment and offline agents, respectively.

- **Step 3: Define PoW**
  Recall from Chapter 6 that PoW is an essential mechanism for the maintenance of tech-driven public goods. Specifically, we are interested in creating club goods, whose regulated access rights avoid the well-known "tragedy of the commons." To regulate access, we exploit the advanced blockchain technology of on-chaining oracles. Recall that on-chaining oracles are instrumental in bringing external off-chain information onto the blockchain in a trustworthy manner. The on-chained information may originate from human users. Hence, on-chaining oracles help tap into human intelligence. As PoW, we define the oracles' contributions to the governance operation of the CPSS via decision-making and knowledge-sharing, which are both instrumental in achieving the specific purpose of CI.

- **Step 4: Design Tokens with Proper Incentive Alignment**
  Most tokens lack proper incentive mechanism design. Recall from Chapter 6 that the use of tokens as incentives lie at the heart of the

DAO and their investigation has started only recently. Importantly, recall that the tokenization process creates tokenized digital twins to coordinate actors and regulate an ecosystem for the pursuit of a desired outcome by including voting rights. The creation of a tokenized digital twin is done via a token contract that incentivizes our defined PoW, involving the following two steps: (i) create digital twin that represents a given asset in the physical or digital world, and (ii) create one or more tokens that assign access rights/permissions of the given physical/digital asset to the blockchain address of the token holder.

- **Step 5: Facilitate Indirect Communication Among DAO Members via Stigmergy and Traces**
  Finally, let the members participating in a given DAO project (i) record their purpose-driven token incentivized activities in ESPN's blockchain-enabled online environment; and (ii) use these blockchain transactions (e.g. deposits) as traces to steer the collective behavior toward higher levels of CI in a stigmergy enhanced Society 5.0.

Figure 8.2 illustrates the functionality of each of these five steps in more detail, including their operational interactions, toward ushering in the future *stigmergic society* that will leverage on time-tested, self-organization mechanisms borrowed from nature. Note that, in doing so, the future stigmergic society follows the guiding principle of *biologization* briefly introduced in Chapter 1 and to be further discussed in Section 8.5.

8.4.2. Implementation

A general definition of human intelligence is the success rate of accomplishing tasks. In our implementation, human intelligence tasks (HITs) are realized by leveraging the image database ImageNet (www.image-net.org) widely used in deep learning research and tokenizing it. Specifically, humans are supposed to discover a hidden reward map consisting of purpose-driven tokens by means of image tagging, which is done by relying on the crowd intelligence of Amazon Mechanical Turk (MTurk) workers and the validation of their answers via a voting-based decision-making blockchain oracle. We measure CI as the ratio of discovered/rewarded number and total number of purpose-driven tokens.

Figure 8.3 depicts the set-up and experimental steps of our implementation in more detail. We developed a JavaScript-based HIT platform to let a human select from 20 ImageNet images as well as add relevant image tagging information and deliver both to the properly configured MTurk and Amazon Web Services (AWS) accounts using an intermediate OOCSI server. The answers provided by MTurk workers to each submitted HIT were evaluated by an on-chaining oracle, which used ERC-20 compliant access right tokens to regulate the voting process and release the purpose-driven tokens

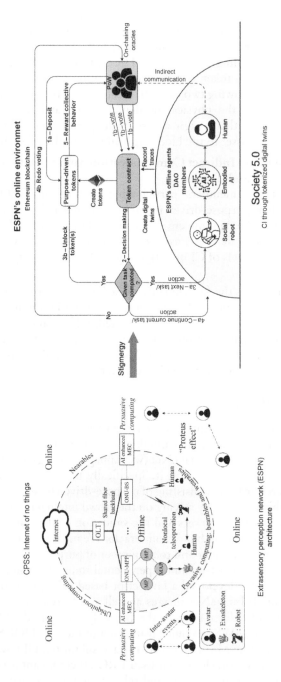

Figure 8.2 Stigmergy enhanced Society 5.0 using tokenized digital twins for advancing collective intelligence (CI) in CPSS.
Source: Beniiche et al. (2022) © 2022 IEEE.

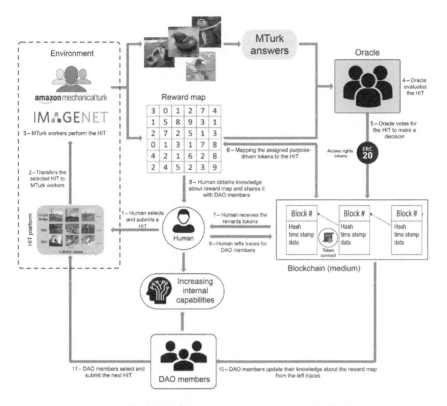

Figure 8.3   Discovery of hidden token reward map through individual or collective ImageNet tagging via Amazon MTurk and on-chaining oracle.
Source: Beniiche et al. (2022) © 2022 IEEE.

assigned to each successfully tagged image. Finally, the human leaves the discovered/rewarded tokens as stigmergic traces on the blockchain to help participating DAO members update their knowledge about the reward map and continue its exploration.

8.4.3. Experimental Results

Figure 8.4 shows the beneficial impact of stigmergy on both CI and internal reward in terms of hidden tokens discovered in the reward map by a DAO with eight members. For comparison, the figure also shows our experimental results without stigmergy, where the DAO members don't benefit from sharing knowledge about the unfolding reward map discovery process.

Given its similarity to decentralized blockchain technology, we adopted stigmergy – a biological self-organization mechanism widely found in social

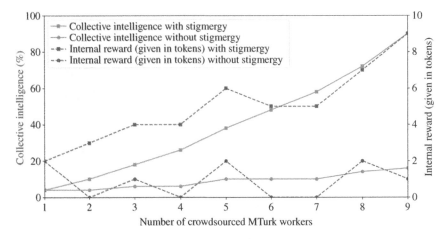

Figure 8.4 Collective intelligence (given in percent) and internal reward (given in tokens) with and without stigmergy vs. number of crowdsourced Amazon MTurk workers.
Source: Beniiche et al. (2022) © 2022 IEEE.

insect societies – for facilitating indirect communication and internal coordination among offline agents via traces (feedback loops) created in a blockchain-based online environment. Our implemented CPSS-based bottom-up token-engineering DAO framework for Society 5.0 is experimentally shown to increase both CI and rewarded purpose-driven tokens by means of stigmergic traces in a blockchain-based online environment involving crowdsourced Amazon MTurk workers and validating on-chaining oracle.

8.5. Biologization: Exiting the Anthropocene and Entering the Symbiocene

Note that human computation via crowdsourcing and biological mechanisms such as stigmergy also hold a promise to benefit from nature's efficiency – a process referred here to as biologization, also known as *bioneering* or *biomimicry* – for "greening" blockchain and future 6G and Next G networks by increasingly using *nature-based solutions* (NbS) that involve working with nature to address societal challenges, e.g. climate change, and provide human well-being.[3]

---

[3]For further information, please visit University of Oxford's NbS Initiative: https://www .naturebasedsolutionsinitiative.org

Recall from Chapter 1 that biologization takes advantage from nature's efficiency for economic purposes – whether they be plants, animals, residues, or natural organisms. It is the guiding principle of the emerging bioeconomy within Industry 5.0. Almost every discipline shares promising interfaces with biology. In the long term, biologization will be just significant as a cross-cutting approach as digitalization already is today. Biologization will pave the way for Industry 5.0 as well as Society 5.0 in the same way as digitalization triggered Industry 4.0. It is also obvious that the two trends – biologization and digitalization – will be mutually beneficial. This mutually beneficial *symbiosis* between biologization and digitalization is anticipated to create exciting opportunities and open up new research avenues in the coming 6G and Next G era.

In a recent essay titled "Exiting the Anthropocene and Entering the Symbiocene," environmental philosopher Glenn Albrecht argues that we must rapidly exit the current human-centered Anthropocene with its nonsustainability and enter the next era in human history named *Symbiocene* (from the Greek *sumbiosis*, companionship or "organisms living together").[4] Exiting the Anthropocene and entering the Symbiocene will be a satisfying experience for most humans. The new era will be characterized by human intelligence and praxis that replicate the symbiotic and mutually reinforcing life-reproducing forms and processes found in living systems. Human development will consist of creative actions that use the very best of biomimicry. However, beyond biomimicry, Albrecht argues that we must also have what he calls *symbiomimicry*. A prime example of symbiomimicry is the so-called *wood-wide-web* – a term coined by the world's leading multidisciplinary science journal Nature. The wood-wide-web regulates forest ecosystems via fungal networks, sometimes referred to as the *Mycelium Internet*, which control nutrient flows to trees that need them most and transfer information and energy from dying species to those that might continue to thrive, thus maintaining the forest as a larger system. These discoveries have transformed our understanding of trees from competitive crusaders of the self to members of a connected, relating, and communicating system. According to Albrecht, these crucially important insights have yet to be incorporated into ecological thinking applied to politics and human societies. As opposed to being anthropocentric, or human-centered, to be *sumbiocentric* means that one is taking into account the centrality of the process of symbiosis in all of our deliberations on human affairs [10].

---

[4]For further information, please visit: https://humansandnature.org/exiting-the-anthropocene-and-entering-the-symbiocene-2021/

References

1. A. Beniiche, S. Rostami, and M. Maier. Society 5.0: internet as if people mattered. *IEEE Wireless Communications*, 29(6), December 2022.
2. D. Ariely. *Predictably Irrational: The Hidden Forces That Shape Our Decisions (Revised and Expanded Edition)*. Harper Perennial, April 2010.
3. S. Voshmgir. *Token Economy: How the Web3 Reinvents the Internet* (Second Edition. BlockchainHub, Berlin, Germany, June 2020.
4. Hitachi-UTokyo Laboratory (H-UTokyo Lab). *Society 5.0: A People-Centric Super-Smart Society*. Springer Open, Singapore, May 2020.
5. Y. Shiroishi, K. Uchiyama, and N. Suzuki. Society 5.0: for human security and well-being. *IEEE Computer*, 51(7):91–95, July 2018.
6. K. Matsuda, S. Uesugi, K. Naruse, and M. Morita. Technologies of Production with Society 5.0. In *Proc., International Conference on Behavior Economic and Socio-Cultural Computing*, pages 1–4, October 2019.
7. P. Chaudhari, R. Utgikar, B. Kelkar, and P. Borse. A Novel Approach: Bioeconomy & Industry 5.0 Enhanced version. In *Proc., IEEE Pune Section International Conference*, pages 1–6, December 2021.
8. S. Wang, W. Ding, J. Li, Y. Yuan, L. Ouyang, and F.-Y. Wang. Decentralized autonomous organizations: concept, model, and applications. *IEEE Transactions on Computational Social Systems*, 6(5):870–878, October 2019.
9. M. E. Gladden. Who will be the members of society 5.0? Towards an anthropology of technologically posthumanized future societies. *Social Sciences*, 8(5):148, 1–39, May 2019.
10. G. A. Albrecht. *Earth Emotions: New Words for a New World*. Cornell University Press, May 2019.

CHAPTER 9

Metahuman: Unleashing the Infinite Potential of Humans

"I have found the missing link between the higher ape and civilized man: It is we."

KONRAD LORENZ
1973 Nobel Laureate in Physiology or Medicine
(1903–1989)

At the end of Chapter 8, we have seen that in the new sustainable era in human history, human development will consist of actions that use the very best of biomimicry as well as the mutually beneficial symbiosis between digitalization and biologization. To unleash the full potential of biologization for the purpose of human development, it is helpful to better understand the biological uniqueness of humans and their possible evolution into future meta-humans with infinite capabilities.

9.1. Becoming Human: The Biological Uniqueness of Humans

Recently, in his award-winning book "Becoming Human," [1] proposed a theory of biological human uniqueness and how we develop the qualities that make us human and become such a distinctive species. In the following, we summarize his major findings from nearly three decades of experimental work with humans and their closest primate relatives.

6G and Onward to Next G: The Road to the Multiverse, First Edition. Martin Maier.
© 2023 The Institute of Electrical and Electronics Engineers, Inc.
Published 2023 by John Wiley & Sons, Inc.

9.1.1. Social Cognition

Mature human thinking is structured by the basic distinction, recognized since the ancient Greeks, between subjective and objective (or appearance and reality, belief and truth, opinion and fact). The distinction derives from the insight that a single individual's subjective perspective on a situation at any given moment may or may not match with the objective situation as it exists independent of this or any other particular perspective. To understand the distinction between subjective and objective, an individual must triangulate on a shared situation with another individual at the same moment: we both see X, but you see it this way, and I see it that way. That is, the participants must come to understand that the two of us are sharing attention to one and the same thing, but at the same time, we each have our own perspective on it.

Great apes and other animal species do not *bifurcate experience* in this way. They take the world as it appears to them, without contrasting it to anything else (objective or otherwise). For instance, a chimpanzee sees a monkey escaping, and he knows that his conspecific sitting next to him sees the monkey escaping also. The conspecific knows the same of his partner. They both are attending to the monkey escaping, and each knows that the other is too. However, they are not jointly attending to it; they are not attending to it together as a "we." Two humans in that same situation could, if so motivated, attend to the monkey escaping together in joint attention. This creates between the two of them a kind of *shared world*, within which they each distinguish their two perspectives. They each also understand that both of their perspectives – that is, their beliefs – on the situation could potentially contrast with an objective (perspectiveless) view of it. Welcome to human reality, humans' unique skills of *social cognition*: socially shared realities and the ability to flexibly manipulate and coordinate different perspectives on aspects of those shared realities (mental coordination). There are two domains of human interaction: (i) social learning (i.e. learning by observing and imitating others) and (ii) communication. These human social/communicative interactions with others create the possibility of new kinds of concepts, including those that depend on an objective perspective.

9.1.2. Coordinated Decision-Making

In collaborative human interactions, the partners coordinate their actions by coming to understand that they are both aiming at the same goal, and that consequently, it is best for them to form a joint goal to pursue it together. Moreover, sometimes they form a joint goal by making together an explicit joint commitment. In both these cases, everything is out in the open: what the other is doing, and what he might be communicating. However, decision theorists have also focused on other situations in which individuals must coordinate not just their actions but also their decisions, and not out in the open but without perceptual access or communication.

The classic example is coordination problems in the game theory sense of the term, giving rise to so-called coordination games, e.g. the stag hunt where individuals cannot see or communicate with each other. In this case, partners need to effect a *meeting of minds* without the support of the perceptual context or communication, which requires one or another form of coordination thinking. In the example of the aforementioned stag hunt, because each partner only wants to go for the stag if the other does as well, the result could easily be paralysis. Chimpanzees are not challenged by the stag hunt because they do not perceive it as dilemma. They never communicate, even though they could have. As a result, their performance went down significantly. Other studies have found a similar lack of communication between apes in collaboration situations in which it would have been beneficial to do so. Children did something that the chimpanzees did not do when there was a barrier: they communicated with one another. They were able to do this both vocally and with gestures by raising above the barrier. This communication enabled them to maintain their same level of success as without the barrier. Children were communicating because they did indeed perceive the dilemma and so helped the partner make his/her decision by advertising their decision, resulting in common-ground understanding of mutual knowledge and recursive understanding. Unlike apes, children were shown to indeed engage in some kind of *recursive mind-reading*.

The close interrelation among coordination, communication, and recursive mind-reading very likely has, at least in part, an evolutionary explanation. Coordinating with others on shared goals in the context of collaborative foraging (see also hunter-gatherer Society 1.0 in Figure 1.4) was very likely the adaptive challenge leading to humans' unique forms of cooperative communication and recursive mind-reading. Humans must align their decision-making with that of a peer partner, which they normally would do by communicating. However, when the normal means of communication are blocked, a human can still coordinate her thinking with a partner by recognizing, recursively, how the partner is thinking about how she is thinking about his thinking, and so forth.

In modern cognitive science, the dominant metaphor for the process of thinking is computation. Nonetheless if we now ask why human thinking and great ape thinking differ in just the ways they do, computation-based theories are impotent: they have no resources for even formulating an evolutionary hypothesis. We might be able to provide a computational answer to the question of how the thinking of an infant differs from that of a six-year-old, but we could not explain how this difference came about unless we simply say it is innate in the genetic blueprint and that is the end of the story. In fact, this is the preferred explanatory strategy for many computational theorists. This explanatory proposal just kicks us back to a version of the evolutionary question: how did the human genetic blueprint get to be the way that it is, such that it now differs in systematic ways from that of other great apes? *We need something more than computation* (to be continued in Section 9.1.4).

9.1.3. Uniquely Human Sociality

The shift to an ultra-cooperative lifestyle during human evolution transformed the nature of human social relationships. Whereas great ape social relations are based mainly on competition and dominance, with a dash of cooperation, early humans began forming with cooperative partners as joint agents "we," comprising "you" and "me" (perspectively defined) as constituents of that "we."

Great apes sometimes collaborate with one another as well, but, tellingly, the most common context is for purposes of competition, as they form coalitions with one another to win fights and other dominance contests. Their "friends" are those with whom they can team up to defeat others. Chimpanzees and other apes thus live their lives embedded in more or less constant competition: they are constantly attempting to outcompete others by outfighting them, outsmarting them, or outfriending them. Great ape collaborators are basically using their partners, in sophisticated ways, as social tools. Although chimpanzee collaboration appears on the surface to be similar to the human version, in reality – in terms of the psychological processes involved – it is less like working together mutualistically and more like individuals using one another to achieve their individual ends. Overall, perhaps with some exceptions, we may say that chimpanzees view others mostly instrumentally: as social obstacles in competition, or as social instruments in collaboration. They operate in *group behavior in I-mode*: they are using one another as social tools.

A basic fact is that human children are more motivated to interact with others in collaborative activities than are great apes. When infants and toddlers interact collaboratively with an adult, they are doing something different. They act together as a joint agent toward a joint goal, which means that they each have their own individual role. Since infant/toddler and adult both take the perspective of the other, and can reverse roles as needed, a new social relationship emerges. The reciprocally defined I–you relation is the foundation for so-called second-personal social relationships, involving two second-personal agents who relate to one another cooperatively, with mutual respect, as equals. It is also noteworthy that children are so trusting of a cooperative motive in others that they engage collaboratively with almost any adult, familiar or novel, as demonstrated in many experimental settings. They form with their partner a joint agent "we" in order to pursue a joint goal, and maintaining this "we" is part of their continuing motivation. Conversely, the reason that great apes do not treat others fairly is that they do not participate in joint intentional collaboration, so they do not form a "we" comprising "I" and "you." As a result, they do not construct a sense for *self-other equivalence* with a partner. The recognition of self-other equivalence is an inescapable fact that characterizes the human condition and uniquely human prosociality.

Collaborating and acting prosocially with a partner eventually scaled up in human evolution to participating in a larger cooperative enterprise known as *culture*. To maintain cooperation within their distinct cultural groups, humans have evolved a unique form of social control in which the group as a whole expresses its collective expectations for individual behavior. These collective expectations are known as social norms, and individuals are normally expected both to conform to them and to enforce such conformity on others. Interestingly, great apes often retaliate against those who have harmed them directly, but they do not punish or intervene against an individual who is harming a third party, even if this is a relative or friend. That is, great apes retaliate against those who harm them, but do not punish actors for harmful acts against third parties. Human social norms, in contrast, take a thoroughly third-party perspective: they apply to everyone in the group alike and should be enforced on everyone alike, even when they themselves have nothing to gain. This suggests that enforcers of social norms are motivated by something more than immediate self-interest. Humans have what we may call a *sense of justice*, which may be defined as a sense of fairness applied not just to individuals but also to the members of a group. While it seems that great apes have no sense of justice, third-party punishment, in one form or another, is a human universal out of some sense of justice to the group at large. The key is that everyone is treated equitably, and when that is violated order must be restored.

9.1.4. How Evolution Makes Us Smarter and More Social

Lev Vygotsky and other sociocultural theorists recognize that much of what makes humans unique is their *sociocultural experiences*. Nevertheless, they fail to recognize humans' biological preparedness specifically for cooperation and culture altogether.

Tomasello [1] developed a modernized neo-Vygotskian theory called *shared intentionality theory*, which is a Vygotskian theory because it is focused not on all of human psychology but only on uniquely human psychology, and it explains uniquely human psychology mainly in terms of the unique forms of sociocultural activity in which individuals engage over the life course. It invokes an evolutionary approach in which individuals are biologically adapted in specific ways for engaging in their species-unique forms of sociocultural activity. Further, it focuses much more on the way that these adaptations facilitate social and mental coordination in such activities as joint attention, collaboration, and cooperative communication.

We may say that, cognitively, the dual-level structure of simultaneous sharedness, i.e. creating *socially shared realities*, and individuality, i.e. individuals' perspectives within those shared realities, characterizes everything from children's pretend play to adults' cultural institutions – that is to say, precisely those sociocultural activities that other great apes can neither create nor understand. The emergence of this dual-level structure of joint intentionality enables humans to have species-unique types of *sociocultural experience*. In other words, one of the main differences between humans and great apes is the creation of socially shared realities and their sociocultural experiences therein. Conceiving of "I" and "you" as equivalent partners within our cooperative "we" characterizes everything in a modern civil society. Again, this is precisely those sociocultural activities that other great apes can neither create nor understand.

It is important to note that a founding principle of evolutionary psychology is that new adaptations only arise in response to specific *ecological challenges*. It is not sufficient to say something like being smart is a generally good thing, so humans evolved to be smart, or being cooperative is a generally good thing, so humans evolved to be cooperative. Evolution does not work that way. Evolution is mostly conservative until a specific adaptive problem presents itself; then those individuals best equipped to solve it have an *adaptive advantage*, and so, because like begets like, the species evolve.

However, the fact that a psychological adaptation is aimed at a specific ecological challenge does not constrain its subsequent application. This applies in particular to the aforementioned shared intentionality of human evolution. Although it was originally selected to deal with a relatively specific set of socioecological challenges presented by the need to cooperate, humans' species-unique capacity for shared intentionality and the experiences that this makes possible empowered individuals to meet those challenges by forming with one another a joint or collective agency. They could now act together and think together, in a sense, as *one*.

According to [1], there have been eight major transitions in the evolution of complexity of living things on planet Earth. Remarkably, in each case, the transition was characterized by the same two fundamental processes: (i) a new form of cooperation with almost *total interdependence* among individuals (be they cells or organisms) that creates a new functional entity, and (ii) a concomitant *new form of communication* to support this cooperation. In this very broad scheme, we may say that shared intentionality represents the ability of human individuals to come together interdependently to act as single agent toward a shared goal or purpose – either jointly between individuals or collectively among the members of a group – maintaining their individuality throughout, and coordinating the process with new forms of cooperative communication, thereby creating a *fundamentally new form of sociality*. In all such cases – at the level of both joint and collective intentionality – the basic structure is a we > me mode of operation in which "we" self-regulate each of us as individuals and their behavioral decision-making, as they gradually become fully fledged persons in a culture.

9.2. Implications of Biological Human Uniqueness for Metaverse

The Metaverse should become a platform that enables human *mass flourishing* – a combination of material well-being and the "good life" in a broader sense. Mass flourishing is a term borrowed from Edmund Phelps's equally named book, in which Phelps, the 2006 Nobel Laureate in Economic Sciences, makes a sweeping new argument about what makes nations prosper. He makes the case that the wellspring of this flourishing is modern values such as the desire to create, explore, and meet challenges. These values fuel the *grassroots dynamism* that is necessary for widespread, indigenous innovation. Phelps notes that most innovation isn't driven by a few isolated visionaries like Henry Ford and Steve Jobs; rather, it is driven by millions of people empowered to think of, develop, and market innumerable new products and processes, and improvements to existing ones. Yet he observes that grassroots dynamism, indigenous innovation, and widespread personal fulfillment weakened decades ago, leading to a narrowed innovation that limits flourishing to a few [2].

Unfortunately, current social media platforms tend to foster the opposite of mass flourishing, as witnessed by today's online tribalism, fake news, or hate speech. In Section 8.3.2, we saw that the tools we have created are ripping apart the social fabric of how society works. It is eroding the core foundation of how people behave by and between each other, leading to widespread *dehumanization* rather than mass flourishing. In order to avoid the manifold dangers, scientists must know what man's nature is and what his built-in purposes are, which are essential to the welfare of society, as postulated by no other than Norbert Wiener, the founder of cybernetics, in his seminal book "The Human Use of Human Beings: Cybernetics and Society" (see Section 8.3.2). Or, as Marshall McLuhan, Canada's eminent media theorist and philosopher, who is credited with predicting the rise of the Internet and phrasing the term "global village", has put it:

> "We shape our tools and then our tools shape us … We become what we behold."

As a consequence, the Metaverse should be created such that it offers users novel tools to shape and behold human co-becoming in a future human-centered Society 5.0 by leveraging on biological human uniqueness. Recall from earlier that humans' unique forms of cooperative communication and recursive mind-reading, especially when the normal means of communication are blocked, were very likely the evolutionary result of the adaptive challenge of coordinating with others on shared goals, as posed by early hunter-gatherer Society 1.0. Also, recall that evolution is mostly conservative until a specific adaptive problem presents itself. Evolutionary psychology states that new adaptations only arise in response to specific ecological challenges. Besides those individuals best equipped to solve it have an adaptive advantage and flourish as empowered individuals by now acting together and thinking together, in a sense, as one. Importantly, recall that the fact that a psychological adaptation is aimed at a specific ecological challenge does not constrain its subsequent application for the sake of human evolution. In response to the global Covid-19 pandemic, virtual experiences and events have skyrocketed in popularity as the online-everything transformation took place. With the mass digital adoption of online social activities accelerated by a global pandemic, we may finally find ourselves on the verge of something big and potentially paradigm-shifting: the Metaverse – offering the adaptive advantage of entering a new post-Covid era of the successor to today's mobile Internet instead of falling back on pre-Covid era business-as-usual practices and their notorious failures, e.g. socioeconomic inequality and environmental unsustainability.

As Society 5.0 envisions the world's first super smart society by using information and communications technologies (ICT) to its fullest, a straightforward yet elegant solution for designing the future Metaverse appears to be to benefit from evolution's capabilities to make us not only *smarter* but also *more social*, two attributes that are front and center in the *super smart Society 5.0* vision. In addition, the process of symbiosis as a major driving force behind evolution should be fully exploited, most notably the mutually beneficial *symbiosis between biologization and digitalization* as the two guiding principles of Society 5.0 and its underlying Industry 5.0 bioeconomy. Recall from Section 9.1.4 that each major transition in the evolution of complexity was characterized by two fundamental processes: (i) a new form of cooperation with almost total interdependence among individuals that creates a new functional entity, and (ii) a concomitant new form of communication to support this cooperation. As a result, a *fundamentally new form of sociality* is created where individuals come together interdependently to act as a single agent toward a *shared goal or purpose*, while maintaining their individuality throughout the process.

We believe that our proposed approach in Chapter 8 of using stigmergy as a biological mechanism for ushering in a future stigmergic society meets most of the aforementioned design goals of the future Metaverse.[1] Our approach records purpose-driven token incentivized activities in a blockchain-enabled online environment and uses these blockchain transactions as traces to steer the collective behavior toward higher levels of collective human intelligence. By the coupling of environmental and social organization, stigmergy's underlying concept of indirect communication is instrumental in maintaining social cohesion and evolution of social life and new self-organizing institutions that operate in a self-sustaining autocatalytic way without requiring any central control entity, whereby individuals work as if they were alone while their collective activities appear to be coordinated, very similar to modern decentralized blockchain networks. The symbiotic convergence of biomimicry (i.e. indirect communication via stigmergy) and advanced digitalization (i.e. purpose-driven tokens recorded in a blockchain-enabled online environment) represents a promising early example of *symbiomimicry* and *sumbiocentric* means to help exit the Anthropocene and enter the Symbiocene in the coming 6G and Next G era, as explained in Section 8.5.

Furthermore, recall from Section 1.3.3 that Society 5.0's inclusion of diverse nonhuman entities – most notably social robots and software artificial intelligence (AI) agents – as participants is nothing new, but instead a return to the unpredictability, wildness, and continual encounters with the other that characterized Societies 1.0 and 2.0, thanks to the prevalence of diverse nonhuman agency resulting from a heavy reliance on animals as key participants in society. We saw in Chapter 7 that the presence of third parties such as blockchain oracles and social robots play an important role in positively changing the economic behavior of individuals and have a potentially greater social impact than monetary incentives, opening up new research avenues for the anticipated paradigm shift from conventional monetary to future nonmonetary economies based on technologies that can measure activities toward human co-becoming, as discussed in the context of robonomics and tokenomics.

These new types of economy anchored in behavioral economics may help overcome the notion of human rationality of individuals driven by self-interest, as traditionally assumed in mainstream economics, where each individual acts independently from each other in market competition (see Section 8.1). As we saw in Section 9.1.3, such a *group behavior in I-mode* is typical for chimpanzees and other apes, who live their lives embedded in more or less constant competition: they are constantly attempting to

---

[1] It is worthwhile to mention that stigmergy is a rather universal biological mechanism in nature, which is not only widely found in social insect societies such as ants and bees but also in neuron communication in the human brain [3].

outcompete others by outfighting them, outsmarting them, or outfriending them. Surprisingly, this offline behavior of apes resembles some of the online activities of humans on social media platforms, e.g. online bullying or unfriending others. By contrast, while great apes do not punish or intervene against an individual who is harming a third party, even if this is a relative or friend, third-party punishment, in one form or another, is a human universal out of some sense of justice to the group at large. Hence, humans act in a *we > me mode of operation* in which "we" self-regulate each of us individuals, whereby everyone is treated equitably, and when that is violated order must be restored. In Chapter 7, we presented some ideas how blockchain oracles can be effectively deployed for the realization of third-party punishment and alternative reward mechanisms in public good games such as the networked N-player trust game, reaching a maximum social efficiency of 100% even without requiring any Ether deposit (see Figure 7.4).

For the Metaverse of the future, it is desirable to further cultivate the aforementioned *we > me mode of operation* of humans by developing the unique qualities that make us human, primarily social cognition via novel *bifurcated experiences*, which characterizes mature human thinking. Recall from Section 9.1.1 that the human reality is determined by humans' unique skills of *social cognition*: socially shared realities and the ability to flexibly manipulate and coordinate different perspectives on aspects of those shared realities, a process also known as mental coordination. Unlike great apes and other animal species, humans are capable of bifurcated experiences by attending to any given situation together as "we," comprising the different perspectives of "you" and "me" as constituents of that "we." Importantly, they each also understand that both of their perspectives – that is, their beliefs rather than truth – on the situation could potentially contrast with an objective (i.e. perspectiveless) view of it. This unique human capability of bifurcated experiences creates between them a kind of *shared world*. This shared world could be the physical world, the realm of physical experiences through the age-old medium of real life. Or, more interestingly, in the case of the future Metaverse, it could be a new kind of shared world arising from the fusion of digital and real worlds.

The creation of novel shared physical and/or digital worlds, in which human social/communicative interactions with others create the possibility of new kinds of concepts, including those that depend on an objective perspective, is of uttermost importance in today's *highly polarized societies* where people appear to live in different worlds, each with its own separate perspective on reality. Clearly, only one of these conflicting human perspectives on reality can be true. Or from a cognitive science perspective, none of them, as explained in more detail in the following text.

9.3. MetaHuman Creator: Building a More Realistic Virtual World

On 11 February 2021, Epic Games announced their *MetaHuman Creator* by revealing a first look at this new browser-based app that will empower anyone to create a bespoke *photorealistic digital human* with a very high fidelity of hair, skin, eyes, teeth, wrinkles, shadows, etc. According to [4], MetaHuman Creator is one step closer to the original Metaverse vision described in Snow Crash (see Chapter 2). The purpose of the MetaHuman Project is to build a more realistic virtual world by crossing the notorious *uncanny valley* of virtual reality (VR), augmented reality (AR), robotics, and photorealistic computer animation.

9.3.1. Crossing the Uncanny Valley

Figure 9.1 illustrates the uncanny valley as the region of negative emotional response toward objects such as humanoid robots that seem almost human. According to Wikipedia, the uncanny valley denotes a dip in the observer's

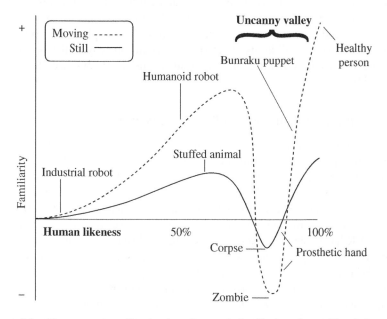

Figure 9.1   The uncanny valley in the observer's familiarity of an object's human likeness. Source: Wikipedia.

familiarity of an object's degree of resemblance to a human being (or sentient being in the case of animals), a relation that otherwise increases with the object's human likeness. Humanoid objects that imperfectly resemble actual human beings provoke uncanny or strangely familiar feelings of uneasiness and revulsion in observers.

Up until now, one of the most arduous tasks in 3D content creation has been constructing truly convincing digital humans. Even the most experienced artists require significant amounts of time, effort, and equipment, just for one character. After decades of research and development, that barrier is being erased through the MetaHuman Creator, which runs in the cloud via Epic Games' Unreal Engine. Once in Unreal Engine, users can animate the digital human asset using a range of performance capture tools. Animations created for one MetaHuman will run on other MetaHumans, enabling users to easily reuse a single performance across multiple Unreal Engine characters or projects. MetaHuman Creator is the next-generation creative tool with all the data coming from real human faces. As a result, users can create characters that are very believable. The development of MetaHumans has attracted increasing attention from major technology companies, given that more and more users are joining Fortnite not to play games, but to socialize. Many players chat with friends on Fortnite, talking about a variety of topics. Instead of seeing Fortnite as a game, they see it as a *social tool* and *social experience*. In this virtual world, people live as in the real world and carry out all kinds of activities. Perhaps one day, real-world events can be moved to Fortnite, such as graduations, birthdays, weddings, and so on [4].

9.3.2. Infinite Reality: Embodied Avatars and Eternal Life

In their book "Infinite Reality: Avatars, Eternal Life, New Worlds, and the Dawn of the Virtual Revolution," which often refers to both novels *Snow Crash* and *Neuromancer* mentioned in Chapter 2, professors Jeremy Bailenson (Stanford University) and Jim Bascovich (University of California, Santa Barbara) provide an account how VR is changing human nature, societies, and cultures as we know them [5]. The two authors make the case that any book about VR has to start with a definition what reality is in the first place. They observe that many scientists, writers, philosophers, and even religious gurus like the Dalai Lama have argued that all perceptions are actually just hallucinations. Ours is not a passive relationship, where reality is and we simply experience it; *reality is, in fact, a product of our minds* – an ever-changing program consisting of a constant stream of perceptions. Besides what they intend to show in their book is how, in many ways, *VR is just an exercise in manipulating perceptions*.

9.3.2.1. Embodied Avatars

VR has accelerated at an astounding pace during the last couple of decades. According to [5], the future of virtual avatars will unfold in the following three evolutionary steps:

1. Avatars will become perceptually indistinguishable from their flesh-and-blood counterparts in terms of how they look, sound, feel, and smell. This is far from an outlandish claim, given how far virtual representations have come in the past decade. Computer graphics used in movies have established that avatars can look just like their owners. As an illustrative example, Bailenson and Bascovich mention the movie *The Curious Case of Benjamin Button*, in which the filmmakers constructed a near-perfect digital replica of Brad Pitt's head and face in order to make him artificially aged. Sounds, smell, feel, and even taste are sure to follow as the years pass, as multisensory extended reality (XR) applications become increasingly reality in future 6G and Next G networks.

2. People's control of their avatars will become automatic, without having to use any type of controller, keyboard, or even vocalization. Avatars will be manipulated via people's everyday actions or thoughts, much like in the movie *The Matrix*, where Neo's avatar interacted with his brain via a neural jack plugged into the back of his head. Note that a similar control was described in William Gibson's *Neuromancer*. As the decades pass, avatars will be controlled as easily as our physical bodies. Recall from Chapter 1 that (wireless) brain–computer interaction is one of the four driving applications behind 6G.

3. Finally, avatars will be embodied and capable of touch, giving rise to future *embodied avatars*. Whether via haptic devices attached to digital simulations, robots tele-operated by owners, or even biological clones that people's minds can control, avatars "will walk among us" and interact with others who are present physically rather than digitally. Today, we see blending of the digital and the physical – for example, kids walking down the streets like zombies (or in the words of Neal Stephenson, author of *Snow Crash*, "Gargoyles") as they text and game within virtual worlds using their smartphones. Thirty years from now, such blending will be routine, and humans and avatars will jostle one another in shopping malls. Note that robonomics, discussed in Chapter 7, and its involved social integration of physical or virtual robotic agents into human society, may be viewed as a harbinger of the dawn of the virtual revolution, which, according to [5], will likely shatter many assumptions about human nature itself, such as *eternal life* as discussed next.

9.3.2.2. Eternal Life: Virtual Immortality

Toward the end of their book, Bailenson and Bascovich argue that avatars can be "more human than human" with regard to their *cyber-immortality*, an idea partially inspired by science fiction written in the 1950s, e.g. Arthur C. Clarke described in his novels how people could live forever by archiving themselves within an advanced computer program. In the physical world, many people do their best to extend longevity with multivitamins, proper diet, and regular trips to the doctor's office. However, if immortality is the goal, then we're currently out of luck biologically. On the other hand, 10 minutes inside a typical VR setup allows digital-tracking equipment to capture literally millions of bytes of data about a person's movements, appearance, and behavior. The amount of data that can be archived about a single person is astronomical, paving the way to cyber-immortality.

It is important to note that the notion of *virtual immortality* differs from the notion of preserving consciousness (to be further explored in Section 9.4). Instead, the idea of virtual immortality is that, with virtual tracking data collected over a long period of time, one can preserve much or even most of people's idiosyncrasies, including a large set of behaviors, attitudes, actions, appearances, etc. One will not be able to relive life through an avatar, but nonetheless, a digital being that looks, talks, gestures, and behaves as they once did can occupy virtual space indefinitely. In this sense, there are two ways of thinking about immortality. One is extending the nature of one's life to be able to continue to enjoy the fruits of living. The other, less experiential, is about preserving one's legacy. Having a version of oneself around forever allows a person to *affect future events* and *shape the experiences of others*. VR will enable nearly everyone to do so.

In VR, people may soon be able to travel back in time to realize the desire of having the opportunity to talk to long-since-passed ancestors. Nevertheless, perhaps even more appealing is to look at the interaction from the perspective of one's grandparents or even great-grandparents. Who wouldn't love to know that long after we are dead, we'd still be able to touch the lives of our progeny generations down the road? It is, in short, a ticket to virtual immortality.

Moreover, virtuality–reality technicians can build a *socially interactive* version of anyone. A 3D digital model provides a near-perfect analog of a person's body and face. Motion-capture technology allows the acquisition of one's gestures. Sooner or later, AI technology will permit implementation of one's personality traits and other psychological idiosyncrasies. Storage of every possible phoneme in one's language, and in one's own voice, will enable one's *eternal avatar* to say things the physical self never even said. Haptic devices capture one's touch – how firm a handshake is, the way one hugs, and how one moves her hands. The overall result will be an avatar that looks and behaves like the person it represents, but can do so even when that person is no longer alive. After one passes on, his grandchildren and their children can enter a *holodeck*, experience his avatar tell a story, give a hug, and provide advice, among many other things. As a consequence, a quite reasonable facsimile of a person's dynamic tendencies can be preserved indefinitely in VR.

9.4. The Case Against Reality: Why Evolution Hid the Truth from Our Eyes

The next-generation Internet referred to as the Web3 will enable the token economy where anyone's contribution is compensated with a token. Recall from Chapter 6 that tokens might be the killer application of blockchain networks and are recognized as one of the main driving forces behind the blockchain-enabled Web3. Importantly, recall that the token economy is not a new thing and has existed long before the emergence of blockchain networks. It has been widely studied in cognitive psychology as a medium of exchange and positive reinforcement method of incentivizing desirable behavior in humans.

In this section, we dig deeper and ask whether cognitive science might also be helpful to better understand the true *nature of reality* grounded in the age-old medium of real life rather than man-made virtual worlds. In particular, cognitive science holds great promise to provide invaluable insights given that the human reality is determined by humans' unique skills of social cognition, as we saw in Section 9.1 in our discussion of the biological uniqueness of humans. Furthermore, social cognition is desirable to be further cultivated toward a *we > me mode of operation* of humans for the future Metaverse (see Section 9.2).

9.4.1. Cognitive Science: From Token Economy to the World as a Network of Conscious Agents

In his critically acclaimed book "The Case Against Reality: Why Evolution Hid the Truth from Our Eyes," the renown cognitive scientist Donald Hoffman from the University of California at Irvine answers the questions whether we can *trust our senses* to tell us the truth and how they can nevertheless be useful even if they are not communicating the truth [6]. He argues that while we should take our perceptions seriously, we should not take them literally.

The world may be viewed as a network of agents. The world has many states that can change and each human perceives the world as an agent, who has a repertoire of certain experiences and actions. The perception and action of an agent are linked in a *perceive-decide-act (PDA) loop*. Based on its current experience, the agent decides whether, and how, to change its current choice of action. The action of the agent changes the state of the world, which in turn changes the experience of the agent. The PDA loop is shaped by an essential feature of evolution – the fitness payoff functions. Each time an agent acts on the world, it changes the state of the world, and reaps a fitness reward (or punishment). Only an agent that acts in ways that reap enough fitness rewards will survive and reproduce. The fitness of an action depends not only on the state of the world but also on the organism (the agent) and its state. Natural selection favors agents with PDA loops properly tuned to fitness. For such an agent, its "perceive" arrow sends it messages about fitness, and its experiences represent these messages about fitness. Importantly, note that the messages and experiences are all about fitness, not about the state of the world. The experiences of the agent become an interface – not perfect, but good enough to guide actions that glean enough fitness points. These fitness points are the reason that perceptions can't show the truth, which is shrouded by a cloud of fitness payoffs.

Hoffman [6] further elaborates on the still largely mysterious concept of consciousness and tries to construct a theory of consciousness, which he calls *conscious realism*, that posits that conscious agents – that is, humans, not objects in space–time – are fundamental, and that the world consists entirely of conscious agents. (Note the human-centeredness of this theory of consciousness!) Specifically, he argues that the external world actually consists entirely of a community or network of conscious agents that enjoy and act on experiences. A simple example consists of only two interacting conscious agents, where the world for each agent is the other agent. How one agent acts will influence how the other perceives. Thus, a single arrow is labeled as both act and perceive. One can consider universes that are more complex, comprising a network of three, four, or even an infinity of agents. The way one agent in a network perceives depends on the way that some other agents act, whereby conscious agents favor interactions that increase mutual information. Besides information, transacted in the currency of conscious experiences, is also the *fungible commodity* of conscious agents, whose central goal is mutual comprehension.

Hoffman notes that many key ideas of conscious realism and the space–time interface theory of perception have appeared in prior sources, from ancient Greek philosophers such as Parmenides, Phythagoras, and Plato through more recent German philosophers such as Leibniz, Kant, and Hegel, and from eastern religions such as Buddhism and Hinduism to mystical strands of Islam, Judaism, and Christianity. In other words, conscious realism is *at once brand new and very ancient*, very similar to the original Metaverse vision (see Section 2.1.4).

9.4.2. The Fitness-Beats-Truth Theorem

In cognitive science, the so-called *fitness-beats-truth* (FBT) Theorem states that evolution is hiding the true nature of reality in space–time from our eyes, i.e. *spacetime is like our own VR* [6]. The FBT Theorem spells out that natural selection does not favor veridical perceptions. It says that natural selection does not shape us to perceive the structure of objective reality, i.e. the true nature of reality, or the truth for short. Instead, it shapes us to perceive fitness payoffs – payoffs that depend not only on objective reality but also on the organism, its state, and its action – and how to get them. As a result, the things that we perceive don't exist independent of our minds.

According to [6], the FBT Theorem has been tested and confirmed in many simulations. They reveal that truth often goes extinct even if fitness is far less complex. Darwin's idea of natural selection entails the FBT Theorem, which in turn entails that the lexicon of our perceptions – including space, time, taste, sound, and smell – cannot describe reality as it is when no one looks. It is not simply that this or that perception is wrong. It is that none of our perceptions could possibly be right. Hoffman concludes that there is an objective reality. However, that reality is utterly unlike our perceptions of objects in space and time.

Then the question remains: what is the objective reality? Hoffman mentions that perhaps our world is a computer simulation, and we are just avatars that haunt it, as discussed in more detail in the subsequent section in the context of the so-called *simulation hypothesis*. However, he makes clear that no class of programs has been found, no scientific theory that starts with neural circuitry has been able to explain the origin of consciousness, and no one has any idea what principle could tie a class of programs to a kind of experience. In short, we have no idea how simulations might conjure up conscious experiences. Simulations run afoul of the hard problem of consciousness: if we assume that the world is a simulation, then the genesis of conscious experiences remains a *mystery*, Hoffman writes.

9.5. Simulated Reality: The Simulation Hypothesis

In this section, we briefly review the simulation hypothesis. In Chapter 2, we have already mentioned it in our brief discussion of *simulated reality*, i.e. reality is nothing more than a computer simulation. The simulation hypothesis is supported by a number of illustrious people such as Larry Page and Elon Musk. Rizwan Virk, an MIT computer scientist, provided a comprehensive description of the simulation hypothesis and its relation to AI, quantum physics, and Eastern mystics in his recently published book "The Simulation Hypothesis," whose major insights we are going to summarize in the following.

According to [7], the simulation hypothesis is first and foremost about computation. It is about all-encompassing simulations like the one depicted in the movie *The Matrix*. The simulation hypothesis actually explains many unexplained questions about our world. This includes some of the paradoxes and unusual aspects of quantum physics as well as the religious views expressed by Eastern mystics and Western religions. Surprisingly, it is the religious views that are closest to the simulation hypothesis: that our physical world is a kind of illusion, populated by conscious beings that exist outside the simulation. As a consequence, one of the most intriguing aspects of the simulation hypothesis may be bridging the gap between two domains of knowledge that rarely overlap, religion and science. Virk argues that the simulation hypothesis may be the only theory that can connect all of these ideas together into a *single, coherent model, or reality*. In short, the simulation hypothesis rationally explains what was previously unexplainable.

It is a popular trend for today's scientists to be atheists, which wasn't always the case. Both Newton and Einstein often talked about God in their writings. Oxford's Nick Bostrom, who first popularized the idea of the simulation hypothesis among academics, has said that speculation of being in a simulation might cause some of them to reconsider their position. Beings that are outside the simulation might seem like "angels" or "gods" to the limited worldview of the beings inside the simulation. In the simulation, someone or something is watching our actions and keeping score – not just what we do but also how it affects other people – in the form of "karmic traces" or "scroll of deeds" on a centralized or decentralized ledger. Interestingly, Virk notes that this concept of "karmic traces" or "scroll of deeds" can be found not only in the Eastern religious traditions but also the Western traditions, including Hinduism and Buddhism ("karma"), Judaism and Christianity ("book of life") as well as Islam ("scroll of deeds"). It is very similar to today's video games and massively multiplayer online role-playing games (MMORPGs), where the gameplay sessions are recorded and played back visually to each player after the game is over to review what has been done well and what hasn't.

Virk concludes that the mystics, who claim to have peeked outside the simulation, may be closer to the *dreamlike nature of reality* than many of the scientists. Mystics of all traditions have told us that what we perceive as reality is actually more like a dream, the grand play that we get caught up in.

9.6. Metareality and Metahuman

To wake up from the dream and bring simulated reality to an end and thereby get access to what Deepak Chopra calls *metareality* involves one thing: shifting from human to *metahuman* [8]. Metahuman helps us harvest peak-experiences that can transform people's lives from the inside out and hint at enormously expanded human potential. The term describes moments when limitations drop away and superb performance happens effortlessly. Metareality is the workshop where consciousness creates everything. It is our source and origin, a field of pure creative potential. It is the quantum field, from which everything in creation springs. Metareality is not perceived by the five senses because it has no shape or location. Yet it is totally accessible, and it offers our only means to escape simulated reality. Being metahuman is like tuning in to the whole radio band instead of one narrow channel. To support this thesis in his book "Metahuman: Unleashing Your Infinite Potential," Chopra cites the German physicist and quantum pioneer Werner Heisenberg as follows: "What we observe is not nature itself, but nature exposed to our method of questioning."

9.6.1. The Human Condition

According to Hannah Arendt, the *human condition* is designated by three fundamental human activities: labor, work, and action [9]. They are fundamental because each corresponds to one of the basic conditions under which life on Earth has been given to man. More specifically, while labor is the activity which corresponds to the biological process of the human body, work (*ergon* in Greek) is the activity which provides an artificial world of things. While labor is the work of our body, *ergon* is the work of hands, e.g. tools.[2] Action, on the other hand, is the only activity that goes on directly between men without the intermediary of things. Action needs the surrounding presence of others no less than work needs the surrounding of nature for its material and of a world in which to place the finished product.

9.6.2. Deus Ex Machina

Interestingly, in her book "The Human Condition," [9] discusses Galileo's discoveries, which undermined traditional ideas about a perfect and unchanging cosmos with the Earth at its center. It was not reason but a man-made instrument, the telescope, which actually changed the physical world view. Man had been deceived so long as he trusted that reality and truth would reveal themselves to his senses and to his reason if only he remained true to what he saw with the eyes of the body and mind. The human eye betrayed man to the extent that so many generations of men were deceived into believing that the sun turns around the Earth. Galileo was a polymath who pioneered the use of the telescope. It was the reading of an instrument that has won a victory over both the senses and the mind. It was the telescope, a work of man's hands, which finally forced nature, or rather the universe, to yield its secrets. Arendt argues that whatever human senses perceive is brought about by invisible, secret forces, and "if through *certain devices, ingenious instruments,*

---

[2]It is interesting to note that the term *technology* is closely related to the work of hands. According to Wikipedia, technology means "science of craft," from Greek *techne*, "art, skill, cunning of hand," and *logia*, "study."

these forces are caught in the act rather than discovered – as an animal is trapped or a thief is caught much against their own will and intentions – it turns out that this tremendously effective Being is of such nature that its disclosures must be illusions and that conclusions drawn from its appearances must be delusions." Arendt concludes that this quality of a *deus ex machina*[3] is the only possible solution to catch the ultimate secrets of Being and transcend appearance beyond all sensual experience, even instrument-aided, in a refound unity of the universe.

Now, wouldn't it be great if the future Metaverse would be a *portal into metareality*, an ingenious "deus ex machina technology" that forces nature to yield its ultimate secrets of Being and catch them in the act rather than discover them? Though man-made, the portal would help us peek into a new kind of shared world arising from the fusion of digital and real worlds beyond our five human senses, enabling us with extrasensory perceptions, sixth-sense experiences, and superhuman capabilities, mentioned in Chapter 2, to create the possibility of new kinds of concepts, including those that depend on an objective reality. Toward this end, the Metaverse should borrow from cognitive science not only the concept of *token economy* (see Chapter 7) but also the concept of *conscious realism*, introduced in Section 9.4, where we saw that our senses cannot be trusted to tell us the truth since physical spacetime is like our own VR. Recall from earlier that conscious realism is at once brand new and very ancient, very similar to the original Metaverse vision.

In the following chapter, we make the proposition that 6G, Next G, and the Metaverse should pave the way for the *peak-experience machine* that helps induce optimal states of consciousness by giving access to the upper range of human experiences, e.g. out-of-body and afterlife experiences, and prioritizing activities over passivities given that people want to do the actions, and not just have the experience of doing them. The peak-experience machine aims at helping people not only have contact with a deeper, non-man-made metareality but also act upon Arendt's aforementioned human condition characterized by fundamental human activities, most notably work (ergon) and action, in their surrounding environment of nature and others.

References

1. M. Tomasello. *Becoming Human: A Theory of Ontogeny.* Harvard University Press, January 2019.

---

[3] The term *deus ex machina* stems from ancient Greek theater, where actors who were playing gods were brought on stage using a machine.

2. E. S. Phelps. *Mass Flourishing: How Grassroots Innovation Created Jobs, Challenge, and Change*. Princeton University Press, August 2013.
3. X. Xu, Z. Zhao, R. Li, and H. Zhang. Brain-inspired stigmergy learning. *IEEE Access*, 7: 54410–54424, April 2019.
4. X. Fang, L. Cai, and G. Wang. MetaHuman Creator: The Starting Point of the Metaverse. In *Proc., International Symposium on Computer Technology and Information Science*, pages 154–157, June 2021.
5. J. Blascovich and J. Bailenson. *Infinite Reality: Avatars, Eternal Life, New Worlds, and the Dawn of the Virtual Revolution*. William Morrow, April 2011.
6. D. D. Hoffman. *The Case Against Reality: Why Evolution Hid The Truth From Our Eyes*. W. W. Norton, August 2019.
7. R. Virk. *The Simulation Hypothesis: An MIT Computer Scientist Shows Why AI, Quantum Physics and Eastern Mystics All Agree We Are In a Video Game*. Bayview Books, March 2019.
8. D. Choprah. *Metahuman: Unleashing Your Infinite Potential*. Harmony, October 2019.
9. H. Arendt. *The Human Condition* (Second Edition). University of Chicago Press, October 2018.

CHAPTER 10

Opportunities vs. Risks

"The trouble with the world is not that people know too little;
it's that they know so many things that just aren't so."

SAMUEL LANGHORNE CLEMENS
a.k.a. Mark Twain
(1835–1910)

In the current human-centered Anthropocene era, today's societies worldwide are undergoing rapid change, both technologically and environmentally. As a result, they are facing a new reality that is characterized by an ever-increasing number of what is often referred to as volatile, uncertain, complex, and ambiguous (VUCA) situations. Within this context of growing uncertainty, this chapter discusses the specific opportunities as well as risks of the Metaverse to become the human-centric approach of choice toward realizing the vision of a super smart Society 5.0 for the twenty-first century.

To rapidly exit the Anthropocene in the coming 6G and Next G era, we have to harmonize our human activities with nature's time-tested self-organization mechanisms found in living systems. More specifically, we have to drive the mutually beneficial symbiosis of digitalization and biologization, the two guiding principles of the future bioeconomy, to enter the next era in human history named Symbiocene, which will foster human development by harvesting satisfying human experiences (see Section 8.5). Recall from Chapter 9 that it is not sufficient to say something like being smart and cooperative is a generally good thing, so humans evolved to be smart and cooperative. Evolution does not work that way. New evolutionary adaptations only arise in response to a specific adaptive problem that presents itself, and those individuals best equipped to solve it have an adaptive advantage. In addition so the human species evolve, whereby an evolutionary

6G and Onward to Next G: The Road to the Multiverse, First Edition. Martin Maier.
© 2023 The Institute of Electrical and Electronics Engineers, Inc.
Published 2023 by John Wiley & Sons, Inc.

adaptation aimed at a specific ecological challenge does not constrain its subsequent application.

With the mass digital adoption of virtual experiences for online meetings and events driven and accelerated by the global Covid-19 pandemic, we finally find ourselves on the verge of something big and potentially paradigm shifting: the online-everything transformation in the form of the emerging Metaverse as the precursor of the Multiverse and the next step after the Internet, similar to how the mobile Internet expanded and enhanced the early Internet in the 1990s and 2000s. As we shall see shortly, however, Internet visions are so abundant that they have even spawned a neologism: *cyberutopianism* – the belief that connecting people through the Internet leads inexorably to global understanding and world peace. It is not enough to be enthusiastic about the possibility of connection across cultures by digital or other means in the anticipation of the rise of a utopian social order.

Instead, *digital cosmopolitanism*, as distinguished from cyberutopianism, requires us to take responsibility for building real and lasting connections in digital space. Data from the Pew Internet and American Life Project suggest that more of our friends on Facebook come from offline associations. Only 7% of Facebook friends in the Pew study were "online-only" relationships; 93% were people whom a Facebook user had met offline [1]. Clearly, these data show that the Internet hasn't made as truly *global citizens* yet. Notwithstanding, global crises such as the current Covid-19 pandemic and impending climate change in the coming years and decades affect practically the entire humanity. As a result, for the first time in human history, we may eventually become global citizens who come together independently to act as a single agent toward a shared goal or purpose in a we > me mode of operation driven by symbiomimicry (see Section 9.2).

The global Covid-19 crisis has been acting like a mirror that makes us see the vulnerabilities and shortcomings of ourselves as individuals and our societies. Throughout human history, crises have created bifurcations where civilizations either regress (e.g. tribalism) or, alternatively, progress by raising their level of complexity through the integration of internal contradictions. In doing so, dichotomies are transcended, and societies are rewired. Each profound crisis is thereby bequeathing a story, a narrative, a code that points far into the future, giving rise to the "new normal," or more precisely, the "new new normal."

10.1. The "New New Normal" and Upcoming T Junction

In his latest book "The Only Game in Town," Mohamed A. El-Erian – the man who originally coined the term the "new normal" after the global financial crisis back in 2008/2009 – argues that now this new normal is getting increasingly exhausted [2]. For those caring to look, signs of stress are multiplying – so much so that the path the global economy is on is likely to end soon, and potentially quite suddenly. As we approach this historic inflection point, unthinkables will become more common. The current path could give way to one of two very different new roads. The first promises higher inclusive growth. In stark contrast, the second would see us mired in even lower growth, rising inequality, political dysfunction, and social tensions.

Fortunately, there is nothing predestined about what will come. El-Erian emphasizes that the road out of the upcoming T-junction can still be influenced in a consequential manner by the choices that we make. Meanwhile, and perhaps more important, *rapid technological innovations* have enabled and empowered individuals like never before. In fact, El-Erian concludes that today, so many more people in so many more places are enabled to connect and participate, and, soon, they will also be able to make a lot more things.

To put this optimism in perspective, it is also important to take critical voices into account to ensure that we have a balanced view on the potentials and limits of future technological innovations. In our introductory discussion of the Metaverse, we have quoted Ethan Zuckerman, former director of the Center for Civic Media at MIT, as a particularly critical voice. Prior to the advent of the Metaverse, Zuckerman has already expressed his critical opinion about the Internet in general in his book "Digital Cosmopolitans: Why We Think the Internet Connects Us, Why It Doesn't, and How to Rewire It" [1]. In it, Zuckerman argues that historically, the arrival of any new technology that has significant power and practical potential always brings with it a wave of visionary enthusiasm that anticipates the rise of a utopian social order. These visions are so abundant that they have even spawned a neologism: cyberutopianism – the belief that connecting people through the Internet leads inexorably to global understanding and world peace.

According to Zuckerman, it is not enough to be enthusiastic about the possibility of connection across cultures, by digital or other means. Digital cosmopolitanism, as distinguished from cyberutopianism, requires us to take responsibility for making these connections real and gain an understanding of what is necessary to build real and lasting connections in digital space. Toward this end, we need to take a close look at the *reality of globalization*, not just the promise, to understand the challenge we face: we have become increasingly dependent on goods and services from other parts of the world, and less informed about the people and cultures who produce them, whereby some globalist aspirations were realized and many were not. Generally speaking, without a way to build personal connections to people from other parts of the world, it is hard for us to take their perspectives and insights seriously. Conversely, encountering the world through *shared interests* is a shortcut to encounters that connect us with other human beings. That step toward finding shared interests is a move toward problem-solving that incorporates diverse and complementary ways of thinking.

The world is currently undergoing a transformation. In his recent 2022 chairman's letter to shareholders of BlackRock, the world's largest asset management corporation, Larry Fink – CEO and chairman of BlackRock – states that the Russian invasion of Ukraine has put an *end to the globalization* we have experienced over the last three decades. The magnitude of Russia's actions will play out for decades to come and mark a turning point in the world order. The ramifications of this war are not limited to Eastern Europe. They are layered on top of a pandemic. According to Fink, the impact will reverberate for decades to come in ways we can't yet predict, though he makes a couple of insightful observations. First, a less discussed aspect of the war is its potential impact on *accelerating digital currencies*. The war will prompt countries to reevaluate their currency dependencies. Even before the war, several governments were looking to play a more active role in digital currencies. Second, during the pandemic, we saw how a *crisis can act as a catalyst for innovation*.

It is important to note that there is a significant difference between globalization and innovation. Thiel and Masters [3] provide invaluable insights into the *future of progress*. That progress can take one of the two forms: (i) horizontal or extensive 1 to n progress, and (ii) vertical or intensive 0 to 1 progress. Horizontal or extensive progress means copying things that work – going from 1 to n. Horizontal progress is easy to imagine because we already know what it looks like. Vertical or intensive progress means doing new things – going from 0 to 1. Vertical progress is harder to imagine because it requires doing something nobody else has ever done. The single word for horizontal progress is globalization – taking things that work somewhere and making them work everywhere. The single word for vertical, 0 to 1 progress is *technology*. According to [3], properly understood, technology is the one way for us to escape competition in a (de)globalized world.

Note that *deglobalization* per se is not a bad thing, if done properly. Balsa-Barreiro et al. [4] have shown that poorly designed networked structures at the global scale add interdependencies that have a hidden downside, as recently witnessed by the elevated risk of contagion during the rapid spread of Covid-19 worldwide and the resultant breakdown of global supply chains. The more connected a system is, the easier it is for errors and unexpected detrimental behaviors to propagate across the system. Interdependencies create new paths for error propagation and may escalate the risks of malfunctions in both frequency and severity. Anomalies do not grow or occur linearly. Their magnitude may explode given the existence of critical masses and tipping points during networked propagation processes. A system with a great number of interdependencies can be extremely vulnerable to malfunctions, even if it does not come from the most important nodes. Similarly, any potential malfunction can effortlessly cascade across the whole system and affect its functional behavior. Whether by over-centralizing or by excessively densifying dependencies, a poor design can lead to the collapse of the whole system. Balsa-Barreiro et al. [4] have shown that in particular centralized and distributed systems behave similarly when the number of connections exceeds a system-dependent threshold number. After the 2008 debt crisis, cascading phenomena have become a hot research topic. Not surprisingly, a substantial part of the literature in economics and finance studies is focused on understanding systemic risks and the stability within the global markets. According to [4], the *structural topology of networks* plays a crucial role for designing methods to create robust networked systems, ranging from the social to the economic and the political.

As shown at the top of Figure 10.1, network topologies may be expressed as varying along a continuum, ranging from a regular network connected to an identical number of neighboring nodes on each side (left) to a fully random network, where connections among nodes are fully random (right).

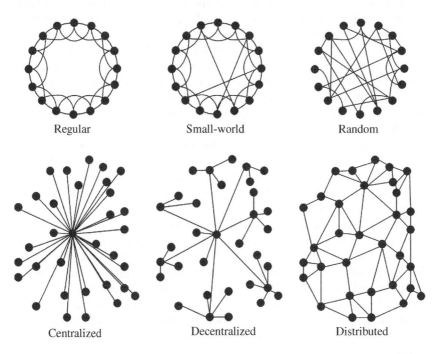

Figure 10.1   Network topologies: regular-to-random continuum (top) and centralized-to-distributed continuum (bottom). Source: Verhagen and Engelen/with permission of Elsevier [5].

At intermediate levels some irregular connections exist between nodes that are far apart (middle) which provide the network with *small-world* properties. Small-world networks look like regular graphs locally, but globally behave like random graphs. The term "small-world" is derived from the experimental work conducted by social psychologist Stanley Milgram in the 1960s, showing that in human social networks two random people are connected by an average of 6 degrees (nodes) of separation, which is also known as "the six degrees of separation theory." Another continuum, that of centralized-to-distributed, is shown in the bottom of Figure 10.1. Nodes in a centralized network are connected via a single central node, a hub (left), and all nodes are involved with the network's interconnectedness to a similar extent in a distributed network (right). Conversely, the intermediate, *decentralized* network topology has many nodes with degree 1, connected to nodes with a higher degree, which in turn are connected to nodes with yet a higher degree. This intermediate case of decentralized networks is similar in topology to *scale-free networks* [5].

According to [5], many real networks have small-world properties and are scale-free. Scale-free networks are commonly found in nature. Scale-free networks have many nodes with few connections and fewer hubs with many connections that make these networks small. Recently, it was reported that the human brain shows small-world, scale-free properties. Small-world, scale-free networks are biologically more economical since they require far less space and energy than a fully random network, resulting in an increased wiring efficiency. Furthermore, Verhagen and Engelen [5] cite Giulio Tononi, one of the most influential neuroscientists at work today, who suggests that the function of the brain is not merely information transmission (as often measured unimodally), but rather the *integration of information* to yield a unified picture of the environment and allow adaptive behavior.[1] This can be achieved by joining smaller complexes with reciprocal connections, leading to higher levels of network complexity. The combined factors of wiring efficiency and network complexity may both have led to the evolutionary development of the highly adaptive brain of humans.

Over the last few decades, our economic systems have been increasingly globalized, thereby promoting a permanent flow of goods and people across borders. The worldwide exchange of goods and people had also hidden downsides, such as potentially irreversible tipping points of environmental pollution and elevated risks of contagious viruses as recently witnessed by the rapid spread of Covid-19 worldwide. Instead of globally trading goods, finding shared interests around the globally uniting issues of Covid-19 and climate change might be a shortcut to encounters that connect us with

---

[1] Giulio Tononi is known for formulating a unified scientific theory that links consciousness to the notion of integrated information, denoted by the Greek letter fi which integrates I (one) and O (zero) into one letter. Tononi argues that consciousness is an evolving, developing, ever-deepening awareness of ourselves in history and culture – that it is everything we have and everything we are [6].

other human beings and their complementary ways of thinking. With the impending end of globalization, the design and deployment of decentralized methods and networked systems will play an increasingly crucial role in improving the resilience of our socioeconomic systems by creating robust networks that exhibit small-world properties of only a few (six on average) degrees of separation. They look like regular graphs locally, but globally behave like random graphs, similar to many scale-free networks commonly found in nature as well as the human brain. By biomimicking the human brain and its aforementioned functions of not mere information transmission but rather the integration of information, future 6G and Next G networks may be instrumental in helping yield a unified picture of the environment that allows adaptive behavior by joining smaller complexes with reciprocal connections, leading to higher levels of network complexity and deeper consciousness and awareness of ourselves. Those small-world reciprocal networks and complexes may help realize the long-term Internet vision of eventually becoming a "global village," a term coined by Canada's eminent media theorist and philosopher in *The Gutenberg Galaxy*, who is credited with predicting the rise of the Internet.

Clearly, the aforementioned discussion shows the critical importance of the emerging Web3 token economy based on purpose-driven tokens and decentralized blockchain technologies. In fact, as we shall see shortly, we move headlong into the paradox of being simultaneously globalized and localized, into a *more decentralized but interconnected world*. In the remainder of this chapter, we highlight some of the opportunities awaiting for the Metaverse, including the emergence of the Global Mind, the social singularity, and the future *peak-experience machine* that democratizes access to the upper range of human experiences and makes them available for the masses in order to foster mass flourishment and unleash the infinite potential of humans. On the flip side, we describe the various pitfalls of the Metaverse, including the risk that it may turn humans into cyborgs or uberworked and underpaid worker bees exploited by an *advanced behavior modification Behaviors of Users Modified, and Made into an Empire for Rent (BUMMER) 2.0 machine*. After weighing both opportunities and risks and outlining possible solutions, we conclude the chapter by envisioning a humanistic setting for the emerging Metaverse in support of *team human* in a future human-centric Society 5.0.

10.2. Opportunities

10.2.1. An Inconvenient Truth for Everything: Earth Inc. vs. Global Mind

Former Vice President and Nobel Peace Prize co-recipient Al Gore's best-selling book "An Inconvenient Truth: The Crisis of Global Warming" exposed the shocking reality of how humankind has aided in the destruction of our planet and the future we face if we do not take action to stop global warming. This book, along with the 2006 companion movie, has gained a great deal of attention, receiving many awards worldwide, including the 79th Academy Award for best documentary feature. A sequel to the book and film, titled "An Inconvenient Sequel: Truth to Power," was released in 2017.

Less public attention was paid to Al Gore's book "The Future: Six Drivers of Global Change," an *inconvenient truth for everything*, a frank and clear-eyed assessment of the emerging forces that are reshaping our world and will continue to do so in the decades to come [7]. In this book, Al Gore argues that ours is a time of revolutionary change that has no precedent in history. While economic globalization has led to the emergence of what he labels *Earth Inc.*, the worldwide digital communications, Internet, and computer revolutions have led to the emergence of the *Global Mind*. The Global Mind links the thoughts and feelings of billions of people and connects intelligent machines, robots, ubiquitous sensors, and databases.

To rethink society, Al Gore advocates to exploit digital tools for growing access to what he calls the Global Mind. The awakening of the Global Mind is disrupting established patterns, creating exciting new opportunities for emergent centers of influence not controlled by elites and the potential for reforms in established dysfunctional behaviors. The outcome of the struggle to shape humanity's future that is now beginning will be determined by a contest between the Global Mind and Earth Inc. In a million theaters of battle, the reform of rules and incentives in markets, political systems, institutions, and societies will succeed or fail depending upon how quickly individuals and groups committed to a sustainable future gain sufficient strength, skill, and resolve by connecting with one another to express and achieve their hopes and dreams for a better world.

But Al Gore warns that attention and focus are diluted on the Internet. The variety of experiences available, the ubiquity of entertainment, and the difficulty in aggregating a critical mass of those committed to change all complicate the use of the Internet as a tool for institutional reform. The addition of three billion people to the global middle class by the middle of this century; however, may be accompanied by new and more forceful demands for democratic reforms of the kind that have so often emerged with the growth of a prosperous and well-educated middle class in so many nations. As the severity of our challenges becomes ever clearer, he is hopeful, even confident, that enough concerned committed individuals and groups will join together in time and self-organize creatively to become a force of reform.

In the digital age, the new digital tools that provide growing access to the Global Mind should be exploited. Internet-empowered precision should be applied to the speedy development of a circular economy, characterized by much higher levels of recycling, reuse, and efficiency in the use of energy and materials. The principles of sustainability should be fully integrated into capitalism. Our current reliance on gross domestic product (GDP) as the compass by which we guide our economic policy choices must be reevaluated. The design of GDP – and the business accounting systems derived from it – is deeply flawed and cannot be safely used as a guide for economic policy decisions. Al Gore concludes that the world community desperately needs leadership that is based on the *deepest human values*.

10.2.2. The Social Singularity

10.2.2.1. Social Physics and Human Hive Mind

In his book "Social Physics," a term originally coined by Auguste Comte, the founder of modern sociology, Alex Pentland, director of MIT's Human Dynamics Laboratory, argues that social interactions (e.g. social learning and social pressure) are the primary forces driving the *evolution of collective intelligence* [8]. According to Pentland, social physics is a new science, offering revolutionary insights into the mysteries of collective intelligence and social influence that can help us design a human-centric society. Social physics is about idea flow, the way human social networks spread ideas and transform those ideas into behaviors. More specifically, by means of *reality mining*, Pentland shows that humans respond much more powerfully to social incentives that involve rewarding others and strengthening the ties that bind than incentives that involve only their own economic self-interest. Collective intelligence emerges through shared learning of surrounding peers and harnessing the power of exposure to cause desirable behavior change and build communities. Further, he observes that most digital media are better at spreading information than spreading new habits due to the fact that they don't convey social signals, i.e. they are socially blind. However, electronic reminders are quite effective in reinforcing social norms learned through face-to-face interactions. He concludes that *humans have more in common with bees* than we like to admit and that future technosocial systems should scale up ancient decision making processes we see in bees.

This conclusion is echoed by Max Borders through his concept of the *social singularity* that defines the point beyond which humanity will operate much like a hive mind (i.e. collective consciousness) [9]. Hives such as swarms of insects are distributed, nonhierarchical systems. There are no managers, no directors, and no assignments from earlier. Planning, such as there is, is carried out in a highly localized fashion by ad hoc teams operating according to their commitment to a mission. Currently, two separate processes are racing forward in time: (i) the technological singularity: machines are getting smarter (e.g. machine learning and artificial intelligence (AI)), and (ii) the social singularity: humans are getting smarter. In fact, he argues that these two separate processes are two aspects of the same underlying process waiting to be woven together toward creating new human-centric industries, where human labor will migrate into more deeply human spheres using the surpluses of the material abundance economy and the assistance from collective intelligence. More and more, we will act like bees to get big things done, whereby humans act as neurons in a *human hive mind* with blockchain technology acting as connective tissue to create virtual pheromone trails, i.e. programmable incentives. According to Borders, we move headlong into the paradox of being simultaneously globalized and localized, into a more decentralized but interconnected world.

10.2.2.2. Ethereum: The World Computer

Recall from Chapter 5 that collective intelligence will also play an important role in the vision of future 6G mobile networks. In contrast to previous generations, 6G will be transformative and will revolutionize the wireless evolution from "connected things" to "connected intelligence" [10]. In fact, according to [11], 6G will play a significant role in advancing Nikola Tesla's prophecy that "when wireless is perfectly applied, the whole Earth will be converted into a *huge brain.*" Toward this end, Strinati et al. [11] argue that 6G will provide an Information and Communications Technologies (ICT) infrastructure that enables human users to perceive themselves as being surrounded by a huge artificial brain offering virtually zero-latency services, unlimited storage, and immense cognitive capabilities. Note that blockchains such as Ethereum are commonly referred to as the *world computer*,[2] which may naturally lend itself to help realize Nikola Tesla's aforementioned prophecy of converting the whole Earth into a huge brain.

10.2.2.3. Extended Stigmergy in Dynamic Media

Stigmergy can be viewed as a unifying concept to study cognition across scales, ranging from intraindividual emergence of mental states in the human brain to intraorganism coordination within living organisms and interindividual emergence of culture in societies [12]. Stigmergic behavior in intelligent systems is a common mechanism to produce cognition in natural societies (e.g. ant colonies) as well as artificial systems (e.g. AI subfield of swarm intelligence). Note that not all self-organized systems need stigmergy. However, all stigmergic systems are self-organized.

Stigmergy is ubiquitous in insect societies that exploit pheromone-based interactions, e.g. foraging large areas around their nest in a parallel and

---

[2]For further information, please visit: https://www.youtube.com/watch?v=j23HnORQXvs™t=28s

robust way. The decision between which trace to follow requires insect-level decision-making. Colony-level decision-making emerges as a result of these local decisions (positive feedback) and pheromone evaporation (negative feedback). Short-term and long-term memories are implemented by insect colonies via the use of multiple pheromones with different evaporation rates. Eventually, spatiotemporal coherence is achieved and the colony migrates to the selected location, implementing a winner-take-all process.

The fact that stigmergy is found also in the human neural system was highlighted for the first time in [12]. Neuronal firing leads to the release of neurotransmitters, which are the molecules that are engaged in the transfer of information from one (presynaptic) neuron to the next one (postsynaptic) neuron in the neural network. Conversely, neuromodulators are molecules that are released from neurons or nearby cells as a consequence of neurotransmitter action that then act back onto the (pre- or postsynaptic) neuron, modulating its activity. A common characteristic of the neuromodulators is that they act at short distances, affecting neurons and synapses located close (i.e. a few cells apart) to their release sites, thus enabling stigmergic interactions between multiple synapses [13]. Note that each synapse, i.e. the connection between each pair of (pre- and postsynaptic) neurons, is not hard-wired. New synaptic connections can form while old ones collapse in a process called structural plasticity of the human brain. Anything from learning new skills to healing after a stroke can result in their restructuring, thereby rewiring the brain.

We have seen that in stigmergic systems, each action of an individual, as a result of her perception of the environment, will change that environment, which will result in a different subsequent perception, in a process that repeats for each and every autonomous individual. Stigmergy is an indirect, mediated mechanism of coordination between actions, in which the trace of an action left on an external medium stimulates the performance of a subsequent action. The traces left in the medium may be interpreted as a form of collective memory. According to [12], the usual assumption is that the medium is passive, meaning that it does not modify the traces. However, in dynamic media, which may be constituted by a population of other types of agents, the medium has some effect over the traces, giving rise to the term *extended stigmergy*.

10.2.3. 6G, Next G, and the Metaverse: Toward the Peak-Experience Machine

Johannes Gutenberg's invention of the printing press in 1450 revolutionized society and heralded 300 years of renaissance. While Gutenberg's invention gave birth to printing, the Internet's full potential still remains to be unleashed in the years to come. According to [14], we do not yet know what the Internet truly is, though its impact is anticipated to be eventually similar or even superior to that of the printing press. Measured in Gutenberg time, we stand today at about the year 1481 with the progression of disruption in society. Note that Luther was born in the year 1483. Hence, the Internet's Martin Luther is yet to come. The emerging Metaverse, the anticipated successor to today's mobile Internet, will be about being inside the Internet (rather than simply looking at it from a phone screen) and producing peak-experiences. This is in contrast to only printing pamphlets about them, as Luther had to do in a pre-Internet era.

10.2.3.1. The Experience Machine

The term *experience machine* was coined by Harvard philosopher Robert Nozick in his 1975 national book award winning bestseller "Anarchy, State, and Utopia" [15]. The experience machine is an imaginative machine that produces favorable sensations by giving users whatever desirable experiences they might want. The experience machine gives users the choice between everyday reality and a presumably preferable simulated reality. Nozick argued that people refuse to be plugged into the experience machine for multiple reasons. Among others, people want to do the *actions* and not just have the experience of doing them. Moreover, plugging into the experience machine limits people to a man-made reality. There is no actual contact with any *deeper reality*, though the experience of it can be simulated. Nozick never tested his claims, but argued that they must naturally be the case.

An interesting approach to turn Nozick's thought experiment of the experience machine into actual reality is the Metaverse. It will be a network of interconnected experiences and devices far beyond mere virtual reality (VR). With the rise of the Metaverse, the Internet will no longer be at arm's length. Instead, it will surround us and will radically reshape society. According to [16], the term "Metaverse" is suddenly everywhere. Apart from Meta, several major players have already embraced the Metaverse, including Microsoft, Apple, Google, Amazon, Samsung, Nintendo, and others.

We are only at the beginning of innovation and experimentation. The Metaverse will put the user first, allowing every member of our society to delve into new realms of possibilities. According to [17], the next phase of the Metaverse is the Multiverse. The various adventures that this place has to offer will surround us both socially and visually.

10.2.3.2. Toward Peak-Experiences

On 25 July 2021, The *New York Times* published an article on *"Facebook's Next Target: The Religious Experience."* In it, Sheryl Sandberg, the company's chief operating officer, is quoted saying that faith organizations and social media are a natural fit because both are about connections. Religion has long been a fundamental way humans have formed community, and now social media companies are stepping into that role. Of course, this is not the first time *The New York Times* writes about religion and science, e.g. Einstein's widely read article on this topic, in which he asserts "that the cosmic religious experience is the strongest and the noblest driving force behind scientific research," even though the churches have always fought against science and persecuted its supporters.[3]

Abraham Maslow, whose famous *hierarchy of needs* is recognized as one of the best-known theories of human motivation, has actually proposed religious experience as a legitimate subject for scientific investigation. In [18], he postulates that an expanded science is needed with larger powers and methods. Such a science includes much that has been called religious. Further, he argues that the nineteenth-century atheist had thrown out the religious questions with the religious answers because he had to reject the religious answers of the churches during the rational "age of philosophers," also known as "enlightenment." But what the more sophisticated scientist is now in the process of learning is that the religious questions themselves are rooted deep in human nature. As a matter of fact, contemporary existential and humanistic psychologists would probably consider a person sick or abnormal in an existential way if he were not concerned with these religious questions. In [18], Maslow introduces the umbrella term of *peak-experiences* to describe transcendent religious experiences, which are now being eagerly investigated by many psychologists. Maslow concludes that not having peak-experiences may be a lower, lesser state, a state in which we are not fully functioning, not at our best, not fully human, not sufficiently integrated. When we are well and healthy and adequately fulfilling the concept of human being, the experiences of transcendence should in principle be commonplace.

The practical importance of those transcendent states in human life and the way we live and work, e.g. Silicon Valley, the Navy SEALs, and maverick scientists, has been recently described in a more comprehensive manner by Kotler and Wheal [19]. The authors describe various clandestine experiments with *ecstatic technologies* that induce these optimal states of consciousness also known as flow, which refers to those "in the zone" moments where focus gets so intense that everything else disappears. The Greeks had a word for this experience: *ecstasis* – the act of "stepping beyond oneself." Plato described ecstasis as an altered state where our normal waking consciousness vanishes completely, replaced by an intense euphoria and a powerful *connection to a greater intelligence*. In ecstatic states, the information we receive can be so novel and intense that it feels like it is coming from

---

[3] Albert Einstein, "Religion and Science," The New York Times, 9 November 1930.

a source outside ourselves. The ecstatic is a language without words that we all speak, a bond linking all of us together. In those VUCA situations of rapid change briefly mentioned at the beginning of this chapter, it is more important than ever for a group to merge action and awareness in order to achieve an astounding level of cognitive dexterity. Through the merger of consciousness and collective awareness, the group stops acting like individuals and starts operating as one – as a single entity, a hive mind. Intelligence gets multiplied, fear divided. The whole is not just greater than the sum of its parts; it is *smarter* and *braver* too. According to [19], researchers found four signature characteristics underneath these optimal states of mental and physical performance: *selflessness, timelessness, effortlessness, and richness*, or *STER* for short. By democratizing access to the upper range of human experience via consciousness-hacking technology, modern-day Gutenbergs and Luthers are taking experiences once reserved for mystics and making them available for the masses. It is interesting to note that at the very end of their book, Kotler and Wheal [19] conclude that those transformative technologies nudge us into clearer self-awareness, providing us a taste of what they call *communitas*, as further discussed in the final Section 10.4.5.

10.2.3.3. Deus Ex Machina Technologies

For illustration, this section highlights a couple of peak-experience examples with potential STER characteristics using some sort of *deus ex machina* technology, which we briefly introduced in Section 9.6.2.

Out-of-Body Experience. Researchers of the Brain Mind Institute at École Polytechnique Fédérale de Lausanne (EPFL), Switzerland, have used a VR headset in combination with a hydraulic motion platform to provide humans with out-of-body experiences. To do so, the researchers developed a VR scenario that visually simulates an out-of-body experience while the user wearing the VR headset is lifted by the hydraulic motion platform. By combining visual and kinesthetic human senses, out-of-body experiences, which reportedly may happen during life-transforming near-death experiences (NDEs), can be artificially created.[4]

Afterlife Experience. Taking it one step further, a heart-wrenching "reunion" between a mother and her deceased daughter aired in a documentary by South Korean broadcaster Munhwa Broadcasting Corporation (MBC). The reunion was made possible by VR technology, which enables the

---

[4]SRF: https://www.youtube.com/watch?v=_ZSXd0KN-0E, minute 34:00-40:20, 22 November 2021 (Online: Accessed on 16 June 2022).

daughter to virtually appear alive. VR haptic gloves together with real-time rendering technology were used to help the mother interact with her daughter, and not just look at her.[5]

The Divine Milieu. G will provide an ICT infrastructure that enables end users to perceive themselves as surrounded by a huge artificial brain offering immense cognitive capabilities [11]. Interestingly, the technology magazine WIRED has published an article about the late Jesuit priest and scientist Teilhard de Chardin's similar vision of a globe, clothing itself with a brain.[6] Chardin, who passed away in 1955, is widely credited with predicting the advent of the Internet. De Chardin [20] elaborated on a new emerging realm, which he named the *divine milieu*, and categorized experiences into activities and passivities. Passivities are things that simply happen to us (e.g. passive consumption of media), whereas activities are efforts we make to promote the development of ourselves and others. Any future peak-experience machine should prioritize activities over passivities. Note that this would encourage people to actually plug into Robert Nozick's aforementioned experience machine, given that people want to do the actions, and not just have the experience of doing them, and have contact with a deeper, nonman-made reality.

10.3. Risks

In this section, we identify some of the challenges the Metaverse might face in the near- to long-term future. These challenges arise not only due to accelerating change of technological capacity but also and more importantly due to style of business plan. Similar to *cybernetic organisms (cyborgs)*, the challenges involve internal neural implants that allow us to enhance and call our intellectual, emotional, and spiritual experiences up at will without requiring people to do any actions, resulting in a rather individualistic passive consumption of the artificially stimulated experiences. Further, the Uberization of everything may exacerbate *platform capitalism* that turns us into uberworked and underpaid worker bees. Given today's social media platforms with their addictive pleasure and reward patterns exploiting a vulnerability in human psychology, there is the risk that the Metaverse may become an advanced *behavior modification machine* on a massive scale.

Throughout the following discussion, we also try to point to proposed solutions to the aforementioned challenges, including a system of *micropayments* suggested already in the 1960s that bear strong resemblance

---

[5]Global News: https://www.youtube.com/watch?v=0p8HZVCZSkc, 14 February 2020 (Online: Accessed on 16 June 2022).
[6]WIRED, "A Globe, Clothing Itself with a Brain," June 1995.

to the emerging Web3 token economy and a more *humanistic setting* for the future of social networking.

10.3.1. Ray Kurzweil's Age of Spiritual Machines

At the beginning of this millennium, in January 2000, Google's former director of engineering Ray Kurzweil published the book "The Age of Spiritual Machines: When Computers Exceed Human Intelligence" [21]. He forecasts that our thinking machines will improve the cost performance of their computing by a factor of 2 every 12 months. That means that the capacity of computing will double 10 times every decade, which is a factor of 1000 every 10 years. So your personal computer will be able to simulate the brain power of a small village by the year 2030, the entire population of the United States by 2048, and a trillion human brains by 2060. If we estimate the human Earth population at 10 billion persons, one penny's worth of computing c. 2099 will have a billion times greater computing capacity than all humans on Earth.

Kurzweil believes that there won't be mortality by the end of the twenty-first century, but not in the sense that we have known it. As we cross the divide to instantiate ourselves into our computational technology, our identity will be based on our evolving mind file. We will be software, not hardware. Today, our software cannot grow. It is stuck in a brain of a mere 100 trillion neural connections and synapses. However, when the hardware is trillions of times more capable, there is no reason for our minds to stay so small. They can and will grow. Just as, today, we don't throw our files away when we change personal computers – we transfer them, at least the ones we want to keep. So, too, we won't throw our mind file away when we periodically port ourselves to the latest, ever more capable, personal computer. Of course, computers won't be the discrete objects they are today. They will be deeply embedded in our bodies, brains, and environment. Our identity and survival will ultimately become independent of the hardware and its survival.

We are discovering that the brain can be directly stimulated to experience a wide variety of feelings that we originally thought could only be gained from actual physical or mental exercise. According to Kurzweil, by the fourth decade of the twenty-first century, we will move to an era of virtual experiences through internal neural implants. With this technology, you will be able to have almost any kind of experience with just about anyone, real or imagined, at any time. You won't be restricted by the limitations of your natural body as you and your partner can take on any virtual physical form. Many new types of experiences will become possible, making real – or at least virtually real – our solitary fantasies. The spiritual experience – a feeling of transcending one's everyday physical and mortal bounds to sense a deeper reality – plays a fundamental role in otherwise disparate religions and philosophies. The notion of the spiritual experience has been reported so consistently throughout history, and in virtually all cultures and religions, that it represents a particularly brilliant flower in the phenomenological garden. With the understanding of our mental processes will come the opportunity to capture our intellectual, emotional, and spiritual experiences, to call them up at will, and to enhance them.

10.3.2. Platform Capitalism: Uberworked and Underpaid

In his book "Uberworked and Underpaid," Scholz [22] states the well-known meme: If the service is free, you are the product. Millions of people are clicking and feeding data on Google, performing searches, and toiling in the capturing apparatus of the knowledge factory. In doing so, it turns us into worker bees for Google, which depends on our pollination. When will the pollination become intolerable? How can the bees be freed from the beekeeper? Scholz makes the important observation that the Internet was started out as a distributed and cooperative network, but then became centralized and corporate. He argues that the Uberization of everything from transportation and food delivery, to medical services and haircuts, has given rise to a *platform capitalism*. What is more, wherever the tech economy is rampant, housing becomes totally unaffordable, leading to gentrification. Scholz demands that we need to reverse that trend.

Toward this end, Scholz advocates the concept of *platform cooperativism*, which builds on the commons and facilitates the cooperative ecosystem. Platform cooperativism consists of the following three parts. First, it is about cloning or creatively altering the technological heart of the sharing economy. Second, platform cooperativism is about solidarity, which is sorely missing in this economy driven by a distributed and sometimes anonymous workforce. Third, platform cooperativism is built on the reframing of concepts like innovation and efficiency with an eye on benefitting all, not just sucking up profits for the few. According to [22], platform cooperativism can invigorate a genuine sharing economy, the solidarity economy. It will not remedy the corrosive effects of capitalism, but it can show that work can be dignifying rather than diminishing for the human experience.

Arguably more interesting, Scholz also discusses the Colin & Collin tax proposal by the French government, which targets Google, Amazon, Apple, and Facebook, asserting that these companies are profiting from the data of the French population without being taxed accordingly. French government asked Nicolas Colin and Pierre Collin to draft a report on the taxation of the digital economy, which they compare to the concept of a carbon tax. The proposed tax system would acknowledge that users in any particular country – France in this case – are part of the operation of companies that offer supposedly free services online. It could be understood as a way to contest the invisible labor of platform capitalism and point to the supporting role of digital laborers who are hidden behind the algorithm. One approach to paying for data labor would be to regulate companies such as Google, Amazon, Apple, and Facebook. This is by no means a new idea in other sectors of the economy such as telecommunications. When industries grow, they often start to get regulated. An alternative approach and highly relevant to this discussion is hypertext pioneer Ted Nelson, who suggested a system of *micro-payments* already in the 1960s, as part of his project Xanadu. This idea of micropayments has come to new prominence because VR pioneer Jaron Lanier promoted it in his book "Who Owns the Future," even though such a micropayment system would require a set-up other than the World Wide Web in its current form. Lanier suggests that while the Internet is poised to rid America's society of its middle class, micropayments could become its savior.

10.3.3. BUMMER Machine 2.0

Jaron Lanier, who coined the term "virtual reality" back in 1987, makes the case that we are constantly prodded by algorithms run by some of the richest corporations in history that have no way of making money other than being paid to manipulate our behavior [23]. What might once have been called advertising must now be understood as continuous *behavior modification* on a titanic scale. Constant, subtle manipulation is unethical, cruel, dangerous, and inhumane. To underline his statement about what he calls behavior modification empires, Lanier quotes Sean Parker, the first president of Facebook, as follows:

> "We need to sort of give you a little dopamine hit every once in a while, because someone liked or commented on a photo or a post or whatever ... It's a social-validation feedback loop ... exactly the kind of thing that a hacker like myself would come up with, because you're exploiting a vulnerability in human psychology ... The inventors, creators – it's me, it's Mark [Zuckerberg], it's Kevin Systrom on Instagram, it's all of these people – understood this consciously. And we did it anyway ... it literally changes your relationship with society, with each other ... It probably interferes with productivity in weird ways. God only knows what it's doing to our children's brains."

Addictive pleasure and reward patterns in the brain – the "little dopamine hit" cited by Sean Parker – are part of the basis of social media addiction, but not the whole story because social media also uses punishment and negative reinforcement. Lanier coins a new acronym to account of the pieces that make up the problem: The *BUMMER* machine. More precisely, Lanier argues that our problem is not the Internet, smartphones, or the art of algorithms. Instead, the problem that has made the world so dark and crazy lately is the BUMMER machine, and the core of the BUMMER machine is not a technology, exactly, but a style of business plan that spews out perverse incentives and corrupts people. The problem isn't any particular technology, but the use of technology to manipulate people, to concentrate power in a way that is so nuts and creepy that it becomes a threat to the survival of civilization. Based on his observation that since social media took off, assholes are having more of a say in the world, it is no surprise that Lanier calls BUMMER platforms "asshole amplification technology," which pushes tribalism. As a consequence, BUMMER platform experiences ricochet between two extremes. Either there is a total shitstorm of assholes or everyone is super careful and artificially nice. Lanier's working hypothesis of the human condition has long been that there is a switch deep in every human personality that can be set in one of two modes. We are like wolves. We can either be solitary or members of a pack of wolves. Lanier calls this inner switch the *Solitary/Pack switch*. When the switch is set to Solitary, we are more free. We are not only cautious but also capable of more joy. We think for ourselves, improvise, and create. We scavenge, hunt, and hide. When the switch is set to Pack, we become obsessed with and controlled by a pecking order. We pounce on those below us and we do our best to flatter and snipe at those above us at the same time. Our peers flicker between ally and enemy so quickly that we cease to perceive them as individuals. The only constant basis of friendship is shared antagonism toward other packs.

According to [23], the BUMMER machine naturally promotes tribalism and is tearing society apart, making social improvement hopeless. Not only is your worldview distorted, but you have less awareness of other people's worldviews. You are banished from the experiences of the other groups being manipulated separately. Their experiences are as opaque to you as the algorithms that are driving your experiences. What is really going on is that we see less than ever before of what others are seeing, so we have less opportunity to understand each other. Lanier recalls when the Internet was supposed to bring about a transparent society. The reverse has happened. The cheerful rhetoric from the BUMMER companies is all about friends and making the world more connected. Moreover, yet research shows a world that is not more connected, but instead suffers from a heightened sense of isolation. Clearly, the Metaverse promoted by Facebook (now renamed Meta) as the next evolutionary step of social media bears the risk of becoming the *BUMMER machine 2.0*.

Despite these dangers of today's online platforms, Lanier remains an optimist about technology. Whereas demonstrating the evil that rules social media business models today, he also envisions a *humanistic setting for social networking* that can direct us toward a richer and fuller way of living and connecting with our world. According to Lanier, collective processes make the best sense when participants are acting as individuals, even though it might sound like a contradiction at first. He concludes that if you design a society to suppress belief in consciousness and experience – to reject any exceptional nature to personhood – then maybe people can become like machines. That is happening with the BUMMER machine. The BUMMER experience is that you're just one lowly cell in the great superorganism of the BUMMER platform. If this new challenge to personhood were only a question of spiritual struggle within each person, then perhaps we could say it is each person's responsibility to deal with it. However, there are profound societal consequences. Lanier claims that spiritual anxiety is a universal key that explains what might otherwise seem like unrelated problems in our world. According to Lanier, these problems might all look different at first, but on close inspection, they are all versions of the same question: what is a person? Or borrowing from Kaplan [24], to translate the question into action: Be a Mensch!

10.4. Team Human: From Communications to Communitas

In this final section of Chapter 10, we present some ideas for the emerging Metaverse, the precursor of the future Multiverse, how to deepen the intended human-centeredness of future 6G and Next G networks by expanding their focus from a narrow communications perspective to a broader *communitas* perspective (a well-known term in anthropology) in support of team human, rather than team machine. Let us explain.

In Section 9.3.2, we saw that reality is, in fact, a product of our minds and VR is just an exercise in manipulating human perceptions. Despite all the recent talk about quantum computers and AI, the human brain remains the most complex machine on the planet. For instance, while common sense is natural for humans, it is far out of reach for AI, let alone self-awareness or consciousness. Unlike the aforementioned simulation hypothesis which is first and foremost about computation, computation-based theories are impotent to serve as an evolutionary explanation of the biological uniqueness of humans, especially their recursive mind-reading-based capability of coordinating their actions and decisions without perceptual access or communication (see Section 9.1.2).

As John Brockman puts it, "AI is today's story – the story behind all other stories. It is the Second Coming and the Apocalypse at the same time: good AI versus evil AI." Recently, Brockman edited a book that provides an unparalleled round-table examination about mind, thinking, AI, and what it means to be human, gathering 25 disparate visions of influential thinkers and their opposing perspectives on AI [25].

10.4.1. Self-Refuting Tech Prophecy

Arguably one of the most interesting perspectives of Brockman's edited volume comes from Harvard psychologist Steven Pinker. In his contribution on tech prophecy and the underappreciated causal power of ideas, Pinker argues that the flaw in today's dystopian prophecies is that they disregard the existence of norms and institutions, or drastically underestimate their causal potency. The result is a technological determinism whose dark predictions are repeatedly refuted by course of events. Since technologically advanced societies have enslaved or annihilated technologically primitive ones, it follows that a supersmart AI would do the same to us. Nevertheless, this scenario is based on a *confusion of intelligence with motivation*. Pinker poses the important question, even if we did invent superhumanly intelligent robots, why would they *want* to enslave their masters or take over the world? There is no law of complex systems that says that intelligent agents must turn into ruthless megalomaniacs.

Fortunately, according to Pinker, today's dystopian prophecies are self-refuting. They depend on the premises that (i) humans are so gifted that they can design an omniscient and omnipotent AI, yet so idiotic that they would give it control of the universe without testing how it works; and (ii) the AI would be so brilliant that it could figure out how to transmute elements and rewire brains, yet so imbecilic that it would wreak havoc based on elementary blunders of misunderstanding.

Pinker emphasizes the fact that AI is like any other technology. It is developed incrementally, designed to satisfy multiple conditions, tested before it is implemented, and constantly tweaked for efficacy and safety. We have become a society obsessed with safety, with fantastic benefits as a result: rates of industrial, domestic, and transportation fatalities have fallen by more than 95% (and often 99%) since their highs in the first half of the twentieth century. Yet tech prophets of malevolent or oblivious AI write as if this momentous transformation never happened and one morning engineers will hand total control of the physical world to untested machines, heedless of the human consequences.

10.4.2. The Human Strategy

In the same volume, Alex Pentland – whose ideas on social physics and the evolution of collective intelligence we briefly discussed in Section 10.2.2.1 – laid out *the human strategy* for coping with AI. Pentland argues that people are scared about AI. Perhaps they should be. However, they need to realize that AI feeds on data. Without data, AI is nothing. You don't have to watch the AI; instead, you should watch what it eats and what it does.

If we have the data that go into and out of each decision, we can easily ask, Is this AI doing things that we as humans believe are ethical? This human-in-the-loop approach is called open algorithms, sometimes referred to as *explainable artificial intelligence* (XAI). XAI is the form of AI that can be understood by humans. It provides humans with the ability to explain how decisions are made by machines, which in turn helps people trust AI instead of feeling like their information is being taken advantage of or used without permission. You get to see what the AIs take as input and what they decide using that input. If you see those two things, you'll know whether they're doing the right thing or the wrong thing. It turns out that is not hard to do. If you control the data, then you control the AI.

More importantly, Pentland argues that the things AI learns don't generalize very well. If an AI sees something it has not seen before, or if the world changes a little bit, the AI is likely to make a horrible mistake. He suggests that, in some ways, it is as far from Norbert Wiener's original notion of cybernetics as you can get, because it is not contextualized; it is a little idiot savant.

However, imagine that you took away those limitations: Imagine that instead of using dumb neurons, you used neurons in which real-world knowledge was embedded. If you make those little neurons smarter, the AI gets smarter. So what would happen if we *replaced the neurons with people*? People have lots of capabilities. They know lots of things about the world; they can perceive things in a broadly competent, human way. What would happen if you had a network of people in which you could reinforce the connections that were helping and minimize the connections that weren't? That begins to sound like a *smart society* or smart organization. The fact that humans have a commonsense understanding that they bring to most problems suggests what Pentland calls the human strategy: Human society is a network just like the neural nets trained for deep learning, but the "neurons" in human society are a lot smarter.

Culture is the result of this sort of *human AI* as applied to human problems. According to Pentland, on the horizon is a next-generation AI vision of how we can make humanity more intelligent by building a human AI. It is precisely at the point of creating *greater societal intelligence*. Fortunately, trust networks (e.g. blockchain networks) give us a path forward to building a society more resistant to echo-chamber problems and developing a new way of establishing social measurements in aid of curing some of the ills we see in society today, including extreme polarization, socioeconomic inequality, truth, and justice. That means we can shape our social networks (and the emerging Metaverse) to work much better and potentially beat all that machine-based AI at its own game as Pentland concludes.

10.4.3. Understanding Thinking: From Silicon Valley and Computation to Human Reasoning and Intelligence

Recently, Adrian Daub, a humanist and professor at Stanford University, published an interesting book titled "What Tech Calls Thinking" to inquire into the intellectual bedrock of Silicon Valley [26]. In his book, Daub locates Silicon Valley's supposedly original, purportedly radical thinking in the ideas of several influential individuals.

One of them is Marshall McLuhan, whom we have already encountered in Chapter 9 in our discussion of the implications of biological human uniqueness for the Metaverse (see Section 9.2). In *Understanding Media*, the 1964 book that made McLuhan a household name, the first chapter is named "the medium is the message," the famous and often poorly understood phrase coined by McLuhan. Daub does an excellent job explaining the meaning of this phrase as follows: the central way in which the medium can be the message has to do with *what it asks us to do*, how it asks us to behave toward it. The medium actively shapes and controls the scale and form of human association. McLuhan thought that electronic media would eventually create a small world of tribal drums, total interdependence, and superimposed coexistence. McLuhan proposed that in fact, history is made by media changing human beings. Shifts in communication change our way of thinking, our way of relating to one another, and our very ways of conceiving of ourselves. At the same time, McLuhan thought that electricity does not centralize, but decentralize. He predicted that unlike newspapers and movies – which draw us into the same streets, theaters, and public squares, where we *become a mass* and *have one unified experience* – electronic media would give us similar experiences but by ourselves, or with our own chosen tribe. McLuhan preached that media were all-pervasive, inescapable, but increasingly decentralized in a "global village." The medium is socially the message, meaning that media remake the way groups and classes of people interact. The means by which information is conveyed does more to our sense and self, to our very personhood, than the information itself. According to McLuhan, to create content is to be distracted. More importantly, to create the platform is to focus on the *true structure of reality*.

Another individual who greatly influenced Silicon Valley's way of thinking was the late René Girard, a French scholar of religion and literature at Stanford University. Daub describes him as a magnetic lecturer and far-ranging thinker who inspired the generation that made Silicon Valley into what it is today, including and above all Peter Thiel, whom we briefly mentioned earlier in our discussion of the future of progress (see Section 10.1). Thiel calls Girard one of the last great generalists who is really interested in everything. Girard believed that he had discovered that all human desire is *mimetic* – anything you desire is a mirror of another person's desire for that same thing. Our desires are not ours, they are born from neither our autonomous whims nor any feature of the desired object. All our desires come out of a *network of copied desires* – we like what others like. Perhaps it is not entirely surprising that someone like Thiel drawn to this mimetic theory saw value in Facebook when Mark Zuckerberg first made his pitch. It is hard not to be struck by the fact that the company

Thiel made most of his money from, Facebook, is all about the *algorithmic* desire for incessant reciprocal rating and awarding of status.

For Thiel, mimetic theory revealed and explained how disturbingly herdlike people become in so many different contexts – something he thinks mimetic theory helped him break out of and manipulate at the same time. By drawing his insight from a fairly niche theory rather than from, say, behavioral psychology, Thiel could reframe what is arguably a cliché as rather knowledge that is generally suppressed and hidden. According to Daub, Girard provided for Thiel a *mystical knowledge* that was, when stripped of its rarefied vocabulary and references, really not that different from the common sense of his particular milieu. He argues that Girard's followers in Silicon Valley get to imagine themselves as keepers of an esoteric knowledge few others possess. At the downside, however, Daub notes that Girard puts his hope for redemption in self-knowledge and seems *genuinely allergic to human community* given the violence he sees at its center (which is diametrically opposed to and the exact opposite of the well-known phenomenon of communitas in social and cultural anthropology, as discussed in more detail in the subsequent section).

In stark contrast to Silicon Valley's adopted mimetic theory and algorithmic notion of human desire is Josef Pieper's classic treatise on leisure as the basis of culture that sets man apart from the other animals [27]. We can read it in the first chapter of Aristotle's *Metaphysics*. The Greek word for leisure is the origin of Latin *scola*, German *Schule*, and English *school*. According to Pieper, the original meaning of the concept of leisure has practically been forgotten in today's leisure-less culture of total work. Not only the Greeks in general – Aristotle no less than Plato – but the great medieval thinkers as well, all held that there was an element in purely receptive looking, not only in sense perception but also in intellectual knowing or, as Heraclitus said, "listening-in to the being of things."

Importantly, Pieper [27] emphasizes the fact that the medieval distinguished between the intellect as *ratio* and the intellect as *intellectus*. Ratio is the power of discursive thought, of searching and re-searching, abstracting, refining, and concluding. Whereas intellectus refers to the ability of simply looking, to which the truth presents itself as a landscape presents itself to the eye. The spiritual knowing power of the human mind, as the ancients understood it, is really two things in one: ratio and intellectus, all knowing involves both. The path of discursive reasoning is accompanied and penetrated by the intellectus' untiring vision, which is not active but passive, or better receptive – a receptively operating power of the intellect. Discursive reasoning and intellective vision are as exclusively opposed to one another as "activity" or "receptivity," or as active effort to receptive absorption. Rather, they are related to each other as effort, on the one hand, and effortless, on the other. In other words, *human reasoning* is comparable to algorithmic processing or computation, where the brain may be viewed as a kind of computer. Conversely, *human intelligence* is more like an antenna receiving information from outside the brain. All human knowing involves both: active reasoning and nonactive, purely receptive intelligence.

The ancients contrasted ratio as the decisively human activity with the intellectus, which had to do with what surpasses human limits. Of course, this superhuman power nevertheless does belong to man. For Pieper, it is essential to the human person to reach beyond the province of the human and into the order of angels, the truly intellectual beings. Human knowing has an element of the nonactive, purely receptive seeing, which is not there in virtue of our humanity as such, but in virtue of a *transcendence* over what is human, but which is really the highest fulfillment of what it is to be human, and is thus truly human after all. Pieper argues that this is necessary not only for the perfection of the individuals themselves but also for the perfection of the whole human community.

10.4.4. Communitas: Fusion of Self and World

10.4.4.1. Transcendent Experiences

According to humanistic psychologist Scott Barry Kaufman, the scientific investigation of *transcendent experiences* represents one of today's most exciting frontiers in the science of well-being. In his latest book "Transcend: The New Science of Self-Actualization," he not only excavates the unfinished elements of Abraham Maslow's famous hierarchy of needs, which we discussed in Section 10.2.3.2 but also updates and extends it with the latest science. Maslow believed that satisfaction of the *metaneeds*, beyond basic safety and actualization of one's own self, are necessary to achieve fullest humanness. More specifically, he explains that there are a variety of transcendent experiences that differ in their intensity and degree of unity with the world [28]. As shown in Figure 10.2, there is a unitary continuum, ranging from the experience of becoming deeply absorbed in an activity, commonly referred to as flow, to experiencing mindfulness, to feeling gratitude for a selfless act of kindness, to merging with a loved one, to experiencing awe at a beautiful natural setting, to being so inspired by something – whether an inspiring role model, virtuoso performance, intellectual idea, or act of moral beauty – that you have a transcendent awakening, all the way up to the great peak and even mystical experiences.

Importantly, Kaufman remarks that while transcendent experiences differ in various ways, they all have in common weakening of the boundaries to connectedness with others, the world, and one's own self. Modern research suggests that the transcendent state of consciousness is related to positive mental health and a greater sense of purpose as well as a motivation for

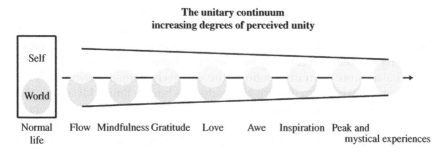

Figure 10.2   Fusion of self and world: unitary continuum with increasing degrees of perceived unity. Source: Kaufman [28].

increased altruism and prosocial behaviors. Kaufman argues that a sort of self-loss occurs during transcendent experiences and that the human capacity to transcend self-interest is the portal to many of life's most cherished experiences. In transcendent moments of self-loss, the experience often feels "realer than real." Self-loss is characterized by high levels of openness to experience. Considering the writings of Maslow and modern psychological research relating to self-actualization and the heights of human nature, Kaufman proposes the following definition of healthy transcendence:

> "Healthy transcendence is an emergent phenomenon resulting from the harmonious integration of one's whole self in the service of cultivating the good society."

Kaufman states that healthy transcendence is a *north star* for all of humanity. In a nutshell: healthy transcendence involves harnessing all that you are in the service of realizing the best version of yourself so you can help raise the bar for the whole of humanity. These "transcenders" reach the highest ceilings of human nature and show us what is possible in all of us and what we could become as a species.

Arguably more interestingly, Kaufman argues that there are promising technologies to *intensify* transcendent experiences. One such technology is VR, a particularly promising tool for generating feelings of awe. Some of the most awe-inspiring experiences are difficult to recreate in laboratory settings but are getting increasingly more realistic in VR technology. VR along with other technologies have the potential to change the course of humanity's future.

10.4.4.2. Communitas: Jumping the Gap into the Future

We have seen in Chapter 1 that from an anthropological perspective Society 5.0 is nothing new, but instead something quite ancient, a return to the unpredictability, wildness, and continual encounters with the other that characterized Societies 1.0 and 2.0, thanks to the prevalence of diverse nonhuman agency resulting from the societies' religious and spiritual dimension (see Section 1.3.3). Also in our discussion of the origins of the Metaverse in Chapter 2, we saw that anthropology plays an important role in understanding the neurological phenomenon of "glossolalia" in many ancient and aboriginal cultures (see Section 2.1.2). Furthermore, recall from Section 8.3.2 that in order to avoid the manifold dangers of cybernetics being exploited to control the human race, Norbert Wiener emphasized the need for the anthropologist, whose insights are essential to the welfare of society.

According to Edith Turner, anthropology concerns itself with humanism. It evokes the human condition in all its messiness, glory, misery, and is thus able to promote cross-cultural understanding and the faculty with which we may grasp what might be called the "ineffable." In her seminal book "Communitas: The Anthropology of Collective Joy," Edith Turner elaborates on the cross-cultural phenomenon of *communitas*, a group's unexpected and unpredictable joy in sharing common experiences, grounded in lived events which may occur as the climax to a process that takes people from violence to shared intimate transcendence. According to [29], communitas should be distinguished from Émile Durkheim's "solidarity," which is a bond between individuals who are collectively in opposition to some other group, leading to an "in-group vs. out-group" opposition. In contrast, in the way communitas unfolds, people's sense is that it is for everybody – humanity, bar none – creating a deep sense of community. Communitas is not limited to any one institution. It does not take sides; it does not rush to "in-group/out-group" competitiveness.

Turner argues that we can find a key to the nature of communitas through the phenomenon of flow, which we have briefly mentioned in our discussion of peak-experiences in Section 10.2.3.2. We often attribute the benefit and wonder of flow to something out there, outside of ourselves. Communitas is concerned with the delicate and permeable energy zone that surrounds people. Turner explains that trying to answer what communitas is, is like trying to locate and hold down an electron. Communitas is activity, not an object or a state. The only way to catch these electrons in the middle of their elusive activity, in process, is to go along with them in the very rush of their impossible energy. One begins to suspect that its coincidence with the mystical side of religions also becomes clearer. Communitas reveals itself through tricks. It derives from nature and in some societies is associated with spirits. There are skills in various cultures aimed to catch the spirit. As a result, communitas does not know limits, it "jumps the gap into the future," as Turner puts it.

Importantly, she notes the joy of pristine small-group communitas in the idea of a kindly *small-scale society* (see also our discussion of the term "small-world" in Section 10.1). Turner can see a current example of this on the Internet, in the exponential effect of the rapid coming together of groups akin to ideas of organic growth. Such interactions show how humans humanize technology and the people's consciousness of their connections with one another. Turner concludes that technology does not necessarily dehumanize humans.

10.4.5. Team Human: Are We in the Midst of the Next Renaissance?

Recall from our aforementioned discussion of the peak-experience machine in Section 10.2.3 that we do not yet know what the Internet truly is, though its impact is anticipated to be eventually similar or even superior to that of the printing press. We argued that the emerging Metaverse, the anticipated successor to today's mobile Internet, will be about being inside the Internet (rather than simply looking at it from a phone screen) and producing peak-experiences. This is in contrast to only printing pamphlets about them, as Martin Luther had to do in a pre-Internet era. Measured in Gutenberg time, we stand today at about the year 1481 with the progression of disruption in society. Note that Luther was born in the year 1483. Hence, the Internet's Luther is yet to come. Gutenberg's invention of the printing press in 1450 revolutionized society and heralded 300 years of *renaissance*. While Gutenberg's invention gave birth to printing, the Internet's full potential still remains to be unleashed in the years to come.

Preeminent digital theorist Douglas Rushkoff argues that our technologies, markets, and cultural institutions – once forces for human connection and expression – now isolate and repress us. In his recently published manifesto *Team Human*, Rushkoff advocates that it is time to remake society together, not as individual players but as the team we actually are [30]. Instead of forging new relationships between people, our digital technologies came to replace them with something else. Our most advanced technologies are not enhancing our connectivity, but thwarting it. Sadly, this has been by design. But that is also why it can be reversed. This is our chance.

Importantly, Rushkoff explains how each new media revolution appears to offer people a new opportunity to wrest the control from an elite few and reestablish the social bonds that media has compromised. But, so far anyway, the people – the masses – have always remained one entire media revolution behind those who would dominate them. For instance, ancient Egypt was organized under the presumption that the pharaoh could hear the words of the gods, as if he were a god himself. The masses, on the other hand, could not hear the gods at all; they could only believe. When writing was finally put in service of religion, only the priests could read the texts and understand the Hebrew or Greek in which they were composed. The masses could hear the Scriptures being read aloud, thus gaining the capability of the prior era – to hear the words of God. However, the priests won the elite capability of literacy. When the printing press emerged in the Renaissance, the people gained the ability to read, but only the king and his chosen allies had the power to produce texts. Likewise, radio and television were controlled by corporations or repressive states. People could only listen or watch. Thanks to online networks, the masses gained the ability to write and publish their own blogs and videos – but this capability, writing, was the one enjoyed by the elites in the prior revolution. Now, the elites had moved up another level and were controlling the software through which all this happened. Developers can produce any app they want, but its operation and distribution are entirely dependent on access to cloud services and devices under the absolute control of just three or four corporations. The apps themselves are merely camouflage for the real activity occurring on these networks: the hoarding of data about all of us by the companies that own the platforms.

Just as with writing and printing, we believe we have been liberated by the new medium into a boundless frontier, even though our newfound abilities are entirely circumscribed by the same old controlling powers. At best, we are settling the wilderness for those who will later monopolize our new world. Rushkoff argues that the problem with media revolutions is that we too easily lose sight of what it is that is truly revolutionary. By focusing on the shiny new toys and ignoring the human empowerment potentiated by these new media – the political and social capabilities they are *retrieving* – we end up surrendering them to the powers that be. Then we and our new inventions become mere instruments for some other agenda. Social phenomena of all sorts undergo this process of hollowing. With digital technology, we too quickly let go of the social and intellectual empowerment offered by these new tools, leaving them to become additional profit centers for the already powerful.

Rushkoff concludes that we are in the midst of a renaissance. The apparent calamity and dismay around us may be less symptoms of a society on the verge of collapse than those of one about to give birth. We may be misinterpreting the natural process of birth as something lethal. One way to evaluate the possibility would be to compare the leaps in art, science, and technology that occurred during the original Renaissance with those we're witnessing today. The Renaissance got its own new media, too: the printing press, which distributed the written word to everyone. We got the computer and the Internet, which distribute the power of publishing to everyone. Most significantly, a renaissance asks us to take a dimensional leap: from 2D to 3D, things to metaphors, or top-down to peer-to-peer. Our renaissance potentially brings us from a world of objects to one of connections and patterns. The world can be understood as a fractal, where each piece reflects the whole. Nothing can be isolated or externalized since it is always part of the larger system. The parallels are abundant. This is our opportunity for renaissance.

However, Rushkoff emphasizes that a renaissance without the *retrieval of lost, essential values* is just another revolution. The beauty of living in a renaissance moment is that we can retrieve what we lost the last time around. The original Renaissance retrieved the values of ancient Greece and Rome. So what values can be retrieved by our renaissance? The values that were lost or repressed during the last one: environmentalism, peer-to-peer economics, and a spirit of mutual aid and community. Just as medieval Europeans retrieved the ancient Greek conception of the individual, we can retrieve the medieval and ancient understandings of the collective. The Renaissance may have brought us from the tribal to the individual, but our current renaissance is bringing us from individualism to something else, aspired to by the distributed economy and blockchain movements – to name just a few. The key to experiencing one's individuality is to perceive the way it is reflected in the whole and, in turn, resonate with something greater than oneself.

According to [30], a renaissance does not mean a return to the past. We don't go back to the Middle Ages. Rather, we bring forward themes and values of previous ages and reinvent them in new forms. Retrieval helps us experience the insight of premodern cultures that nothing is absolutely new. Everything is renewal.

References

1. E. Zuckerman. *Digital Cosmopolitans: Why We Think the Internet Connects Us, Why It Doesn't, and How to Rewire It*. W. W. Norton, November 2014.

2. M. A. El-Erian. *The Only Game in Town: Central Banks, Instability, and Recovering From Another Collapse*. Random House Trade, May 2017.

3. P. Thiel and B. Masters. *Zero to One: Notes on Startups, or How to Build the Future*. Currency, September 2014.

4. J. Balsa-Barreiro, A. Vié, A. J. Morales, and M. Cebrián. Deglobalization in a hyper-connected world. *Palgrave Communications*, 6(28):1–4, February 2020.

5. J. V. Verhagen and L. Engelen. The neurocognitive bases of human multimodal food perception: sensory integration. *Neuroscience & Biobehavioral Reviews*, 30(5):613–650, February 2006.

6. G. Tononi. *Phi: A Voyage from the Brain to the Soul*. Pantheon, August 2012.

7. A. Gore. *The Future: Six Drivers of Global Change*. Random House, January 2013.

8. A. Pentland. *Social Physics: How Good Ideas Spread - The Lessons From A New Science*. Penguin Press, February 2014.

9. M. Borders. *The Social Singularity: A Decentralist Manifesto*. Social Evolution, June 2018.

10. K. B. Letaief, W. Chen, Y. Shi, J. Zhang, and Y. A. Zhang. The roadmap to 6G: AI empowered wireless networks. *IEEE Communications Magazine*, 57(8):84–90, August 2019.

11. E. C. Strinati, S. Barbarossa, J. L. Gonzalez-Jimenez, D. Kténas, N. Cassiau, L. Maret, and C. Dehos. 6G: The next frontier: from holographic messaging to artificial intelligence using subterahertz and visible light communication. *IEEE Vehicular Technology Magazine*, 14(3):42–50, September 2019.

12. L. Correia, A. M. Sabasti ao, and P. Santana. On the role of stigmergy in cognition. *Progress in Artificial Intelligence*, 6(1):79–86, March 2017.

13. X. Xu, Z. Zhao, R. Li, and H. Zhang. Brain-inspired stigmergy learning. *IEEE Access*, 7:54410–54424, April 2019.

14. J. Jarvis. *Gutenberg the Geek*. Amazon Digital Services, February 2012.

15. R. Nozick. *Anarchy, Sate, and Utopia*. Basic Books, 1974.

16. M. Ball. *The Metaverse: And How It Will Revolutionize Everything*. Liveright, July 2022.

17. R. Higgins. *METAVERSE: A Definitive Beginners Guide to Metaverse Technology and How You Can Invest in Related Cryptocurrencies, NFTs, Top Metaverse Tokens, Games, And Digital Real Estate*. November 2021.

18. A. H. Maslow. *Religions, Values, and Peak-Experiences*. Penguin Books, April 1994.

19. S. Kotler and J. Wheal. *Stealing Fire: How Silicon Valley, the Navy SEALs, and Maverick Scientists Are Revolutionizing the Way We Live and Work*. Dey Street Books, May 2018.

20. T. De Chardin. *The Divine Milieu Explained: A Spirituality for the 21st Century*. Paulist Press, November 2007.

21. R. Kurzweil. *The Age of Spiritual Machines: When Computers Exceed Human Intelligence*. Penguin Books, January 2000.

22. T. Scholz. *Uberworked and Underpaid: How Workers Are Disrupting the Digital Economy*. Polity, December 2016.

23. J. Lanier. *Ten Arguments For Deleting Your Social Media Accounts Right Now*. Henry Holt and Company, May 2018.

24. M. Kaplan. *Be a Mensch: Why Good Character Is the Key to a Life of Happiness, Health, Wealth, and Love*. Gefen Publishing House, October 2009.

25. J. Brockman. *Possible Minds: 25 Ways Of Looking At AI*. Penguin Publishing Group, February 2019.

26. A. Daub. *What Tech Calls Thinking: An Inquiry into the Intellectual Bedrock of Silicon Valley*. FSG Originals, October 2020.

27. J. Pieper. *Leisure: The Basis of Culture*. St. Augustine's Press, 1998.

28. S. B. Kaufman. *Transcend: The New Science of Self-Actualization*. Penguin Publishing Group, May 2020.

29. E. Turner. *Communitas: The Anthropology of Collective Joy*. Palgrave Macmillan, December 2011.

30. D. Rushkoff. *Team Human*. W. W. Norton, January 2019.

CHAPTER 11

Conclusion and Outlook

"When wireless is perfectly applied, the whole Earth will be converted into a huge brain."

"My brain is only a receiver, in the Universe, there is a core from which we obtain knowledge, strength, and inspiration. I have not penetrated into the secrets of this core, but I know that it exists."

NIKOLA TESLA
(1856–1943)

11.1. Today's Life in "Parallel Universes"

In the current human-centered Anthropocene era, today's societies worldwide are undergoing rapid change, both technologically and environmentally. As a result of growing uncertainty, they are facing a new reality that is characterized by an ever-increasing number of so-called volatile, uncertain, complex, and ambiguous (VUCA) situations. The global Covid-19 crisis has been acting like a mirror that makes us see the vulnerabilities and shortcomings of ourselves as individuals and our societies. Throughout human history, crises have created bifurcations where civilizations either regress (e.g. tribalism) or, alternatively, progress by raising their level of complexity through the integration of internal contradictions. In doing so, crises act as a catalyst for innovation in that dichotomies are transcended and societies are rewired. Each profound crisis is thereby bequeathing a story, a narrative, a code that points

6G and Onward to Next G: The Road to the Multiverse, First Edition. Martin Maier.
© 2023 The Institute of Electrical and Electronics Engineers, Inc.
Published 2023 by John Wiley & Sons, Inc.

far into the future, giving rise to the "new normal," or more precisely, the "new new normal" – the upcoming T-junction, a historic inflection point – as the "new normal" after the global financial crisis back in 2008/2009 is getting increasingly exhausted. Properly understood, technology is the one way for us to escape competition in a (de)globalized world.

The creation of novel shared physical and/or digital worlds, in which human social and communicative interactions with others create the possibility of new kinds of concepts, including those that depend on an objective perspective, is of uttermost importance in today's highly polarized societies where people appear to live in different worlds or, as the saying goes, in "parallel universes," each with its own separate perspective on reality. Clearly, only one of these conflicting human perspectives on reality can be true, or more technically, from a cognitive science perspective, none of them.

Unfortunately, current social media platforms tend to foster the opposite of human mass flourishing, as witnessed by today's online tribalism, fake news, or hate speech. The tools we have created are ripping apart the social fabric of how society works. It is eroding the core foundation of how people behave by and between each other, leading to widespread dehumanization. Internet visions are so abundant that they have even spawned a neologism: cyberutopianism. It is not enough to be enthusiastic about the possibility of connection across cultures by digital or other means in the anticipation of the rise of a utopian social order. As the anticipated successor of today's mobile Internet and precursor of the future Multiverse, the Metaverse might face various pitfalls in the near- to long-term future, including the risk that it may turn humans into cyborgs with internal neural implants or uberworked and underpaid worker bees exploited by an advanced behavior of users modified, and made into an empire for rent (BUMMER) 2.0 machine using addictive pleasure and reward patterns to exploit a vulnerability in human psychology on a massive scale. The BUMMER 2.0 machine naturally promotes tribalism and tears society apart, making social improvement hopeless, because one is banished from the experiences of the other groups being manipulated separately.

Despite these risks, one should remain an optimist about technology since there exist promising solutions to tackle the involved challenges. Among others, a system of micropayments was suggested already in the 1960s, which bears strong resemblance to the emerging Web3 token economy. While the current Internet is poised to rid society of its middle class, micro-payments could become its savior. Moreover, a more humanistic setting envisioned for the emerging Metaverse holds great promise to support team human in a future human-centric Society 5.0 by directing us toward a richer and fuller way of living and connecting with our world.

11.2. The Road Ahead to the Future Multiverse: 6G, Next G, and the Metaverse

The Metaverse will be the precursor of the future Multiverse. It might be viewed as the next step after the Internet, similar to how the mobile Internet expanded and enhanced the early Internet in the 1990s and 2000s. It will surround us both socially and visually, spanning a wide range of interconnected platforms as well as the digital and physical worlds underpinned by decentralized Web3 technology. The Web3 will enable the token economy where anyone's contribution is compensated with a token.

The term "token economy" is far from novel. In cognitive psychology, it has been widely studied as a medium of exchange and positive reinforcement method for establishing desirable human behavior. Unlike coins, tokens may serve a wide range of different nonmonetary purposes. Such purpose-driven tokens are instrumental in incentivizing an autonomous group of individuals to collaborate and contribute to a common goal and the creation of tech-driven public goods. The exploration of tokens, in particular different types and roles, is still in the very early stages. Tokens might be the killer application of blockchain networks. The use of tokens could also enable completely new use cases, business models (e.g. play-to-earn/own tokens), and asset types that were not economically feasible before, and potentially enable completely new value-creation models.

11.2.1. Metaverse: Mirror of Real World Rather than Virtual Escape Hatch

The Metaverse holds promise to open up new avenues for earning a living and compensating for a broad and diversified spectrum of previously unrewarded creative activity. The Metaverse is all about shared experiences, a virtual economy, and activities that combine the real and virtual worlds. Unlike cyberspace that resides entirely in virtuality, the Metaverse aims at connecting virtuality with reality, making it possible for people and other sentient beings, intelligent mobile robots, as well as software artificial intelligence (AI) agents to communicate and interact in shared environments.

At the downside, we have seen in Chapter 1 that there have been some critical voices recently surfacing about the lack of compelling use cases and potential pitfalls of the Metaverse, even though they might be seen as rejecting the Metaverse in the same way many dismissed the Internet in its early days of 1990s. Perhaps most notably, the Metaverse shouldn't be about building perfect virtual escape hatches – it is about holding a mirror to our own broken, shared world.

According to [1], life in the Metaverse defies and challenges core notions of what it is to be human. Winters emphasizes that it is important to understand the original vision and origins of the Metaverse, which connects with reality and where anyone can have a persistent presence as an extension of their physical self or company. In fact, he argues that it is perhaps from Neal Stephenson where we find the best insight into what the eventual Metaverse might look like. As was the case in Snow Crash, the Metaverse must have the technical capacity to support permanent asset ownership, live interactions, and a multifaceted economy. Ethereum helps keep track of changes in ownership via tokens. He notes that derived from the Old English term *tācen* (meaning sign or symbol, and thus bearing some similarity to the concept of stigmergy), a token is a special-purpose store of value, such as a laundry, arcade, or transportation token. Tokens can be programmed to serve many functions. A token can concurrently serve as proof of ownership over a virtual asset (e.g. virtual land) and as a right to vote in that same virtual world (e.g. governance).

Contrary to popular belief, Winters points out that the concept of virtual or digital real estate is not as novel as most people believe. The reason that virtual property is suddenly gaining more attention now is largely due to the addition of blockchain technology, which verifies digital ownership, builds trust over ownership among all participants in the virtual economy, and helps encourage an active community.

Arguably more importantly, he notes that in developed economies, citizens are often resistant toward the concept of the Metaverse because they value the in-person relationships and experiences already available to them. Moreover, convincing someone to change her mind and lead peers to think differently is like asking them to change tribes. By contrast, according to Winters, the narrative is somewhat different in the Global South, where for some people, an alternative world offers more opportunities and new experiences than their own world. For instance, during the Covid-19 pandemic, there were leagues of blockchain gamers in the Philippines earning US$500 in tokens per month playing Axie Infinity. This is a life-changing source of income for these gamers. Clearly, low-income individuals in the Global North could equally reap such benefits. The Metaverse therefore has great potential to tackle the rich–poor gap not only between countries but also within countries – both developing and developed ones.

We have seen in Chapter 2 that the Metaverse is based on the social value of Generation Z that online and offline selves are not different. We have also seen that originally the term "avatar" means an alter ego that has descended to the earth. It will be necessary to build the Metaverse with a worldview and ethical consciousness. Importantly, the virtual environment, in which people live using avatars, differs from how society currently operates. Toward this end, cross-disciplinary research is necessary, involving cognitive science, social sciences, psychology, and economics. As discussed in Chapter 2, neuroscience and psychological approaches should be used to better understand humans and create and maintain a deeper Metaverse. In particular, we have elaborated on the central role of cyber-physical-social system (CPSS) in the emerging Metaverse for creating a better world for humanity and sustainability. In fact, a concrete Metaverse is just a specific CPSS that needs the support of new technologies developed in blockchains, decentralized autonomous organizations (DAOs), smart contracts, and Web3 as well as a new philosophy of intelligence to transform our worlds into smart societies based on human, artificial, natural, and organizational intelligence.

11.2.2. Overcoming 6G and Next G Blind Spots

We have discussed intriguing 6G visions that foresee the emergence of new themes. Among others, wearable devices are anticipated to fold into our surroundings, back into our physical and biological environments. Unifying experiences across the physical, biological, and digital worlds will enable human transformations, aiming at transforming the behavior of humans through social influence by means of user-intended services via integrated ubiquitous, pervasive, and persuasive computing. Further, the creation of virtual worlds will create a mixed-reality, super-physical world that enables new super-human capabilities. The rise of a new regime will connect all humans and machines into a global matrix, which some call the global mind or world brain, leveraging on the collective intelligence of all humans combined with the collective behavior of all machines, plus the intelligence of nature, plus whatever behavior emerges from this whole.

This book aimed at weaving the aforementioned themes carefully together in future 6G and Next G networks and the enhanced services they offer to disruptive applications in order to enable peak-experiences and human transformations. We paid particular attention to the fusion of digital and real worlds across all dimensions in the recently emerging Metaverse and the closely related Multiverse and its different types of reality created and delivered by nontraditional converged service platforms, where developers do not hesitate to use technologies from as many disciplines as possible, including but not limited to technological disciplines. Importantly, this book aimed at providing the reader with new complementary material, putting a particular focus on 6G and Next G networks in the context of the emerging Metaverse as the successor of today's mobile Internet and precursor of tomorrow's Multiverse. Specifically, we hope that this book is instrumental in helping the reader find and overcome the following 6G and Next G blind spots:

- **Overcoming Blind Spot #1: Next G Expands Incremental 6G = 5G + 1G Mindset of Past Generations of Mobile Networks**
  The Metaverse represents one of the important long-term Next G research objectives along with other important research topics such as quantum networks and holographic calls. 6G paradigm shifts are anticipated to move to a cyber-physical continuum between the connected physical world of senses, actions, and experiences and its programmable digital representation, enabling services beyond communication for a broad range of purposes. To make sense of the future virtual economy enabled by the convergence of wireless and Internet technologies, we need to envision 6G not only from technological but also business and societal perspectives in a multidisciplinary way. In particular, the innate and pervasive integration of blockchain in the 6G ecosystem goes well beyond the 5G case. As a result, it is essential to explore what more the blockchain technology can bring to the 6G realm. Clearly, given these challenges, 6G should be more than only another cellular technology upgrade. Instead, 6G should go beyond continuing the linear incremental thinking 6G = 5G + 1G of past generations of mobile networks. We have seen that Next G research is not just 6G. More specifically, Next G research is independent from the various efforts carried out by the different 6G standard development organizations (SDOs). Hence, Next G includes but is not limited to the specific key performance indicator (KPI) requirements and topics of interest addressed by 6G SDOs. Toward this end, the Next G Alliance has identified six audacious goals, including enhanced 6G Digital World Experiences (DWEs) that transform human interactions across physical, digital, and biological worlds to yield human and machine experiences unthinkable with previous generations of mobile networks. Furthermore, the Next G Alliance emphasizes that sustainability must be at the forefront. According to the Next G Alliance, there is a unique opportunity to consider how to address societal and economic challenges as well as the interdependencies between human and technological evolution. Importantly, there is a symbiotic relationship between technology and a population's societal and economic needs. As technology shapes human behavior and lifestyles, those needs shape technological evolution.
- **Overcoming Blind Spot #2: Network Intelligentization – From AI-Native 6G Networks to Native Edge AI Mimicking Nature Through Brain-Inspired Stigmergy**

6G will be transformative and will revolutionize the wireless evolution from connected things to connected intelligence. 6G will play a significant role in advancing Nikola Tesla's prophecy that "when wireless is perfectly applied, the whole Earth will be converted into a huge brain." In AI-native 6G networks, AI challenges for 6G include the developments of accretionary learning and meta learning. Meta learning, also known as learning-to-learn, has become an active research area recently. The theory of accretionary learning for human beings was proposed by cognitive psychologists over 40 years ago. Adding "intelligence genes" into the network to form intelligence and self-evolution capabilities give rise to intelligence-endogeneous networks (IENs) with self-evolution capabilities, which represent an interesting example of biologization. To further imbue native intelligence, mimicking nature for realizing innovative edge AI-empowered future networks that provide connected intelligence has received an increasing amount of attention recently. Specifically, brain-inspired stigmergy-based federated collective intelligence mechanisms hold promise to accomplish multiagent tasks through simple indirect communication. Edge AI serves as a distributed neural network to imbue connected intelligence in 6G, thereby enabling intelligent and seamless interactions among the human world, physical world, and digital world, acting pretty much like the global mind or world brain.

■ **Overcoming Blind Spot #3: Current Confusion and Uncertainty About Metaverse Is Reduced by Finding Best Insights in Its Origins**
At the time of writing, the term "Metaverse" has no consensus definition or consistent description. Most industry leaders define it in the manner that fits their own worldviews and/or the capabilities of their companies. Notwithstanding, it is fair to assume that there can be only one Metaverse – just as there is "the Internet," not "an Internet" or "the Internets." Recall from above that it is perhaps from Neal Stephenson where we find the best insight into what the eventual Metaverse might look like. In Chapter 2, we have set the stage by describing Stephenson's original vision and the origins of the Metaverse in a comprehensive manner. Stephenson's neologism comes from the Greek prefix "meta" and the stem "verse," a back-formation of the word "universe." For short, in English, "meta" roughly translates to "beyond" or "which transcends" the word that follows. Simply put, the Metaverse is envisioned as a parallel plane for human leisure, labor, and existence more broadly. We have seen that Snow Crash is not only a virus but also a drug and, somewhat surprisingly, even a religion received by receptors built into our brain

cells and expressed thru the language of Nature. In Stephenson's Metaverse, the key realization was that there is no difference between modern culture and Sumerian. The Metaverse is at once brand new and very ancient, getting us to observe the birth of a new religion. Neurolinguistic hacking was developed as a new powerful technology to broadcast instructions directly into people's brainstems. They will act out the instructions as though they have been programmed by a digital metavirus. There is no way to stop the binary virus. However, there is an antidote to it. It would jam the people's mother-tongue neurons and prevent them from being programmed.

- **Overcoming Blind Spot #4: Beyond the Metavervse Origins**
The Multiverse goes beyond the Metaverse in that it creates third spaces that involve experience realms other than reality and virtuality. In an experience economy, people desire experiences that engage them in inherently personal ways, emotionally, physically, intellectually, and spiritually. Value will be determined more by how time is spent and less by how money is spent. The Multiverse spans the entire reality–virtuality continuum. Expanding the Multiverse realms of experience outward and encompassing ever more possibility are instrumental in creating deeper and more intense experiences that eventually help enable the transformation of humans and reflect on what it means to be human? As discussed in Chapter 3, beyond the real and the virtual lays another realm: the Eternal, which lies beyond both universe and Multiverse, both reality and virtuality. Moreover, that existence speaks to the ultimate purpose of everything we do through our acts and ourselves as the means to reach beyond, to find the Infinite.

- **Overcoming Blind Spot #5: Exodus to a Flourishing Metaverse Economy, Culture, and Community**
The extent to which the Metaverse succeeds will depend on whether it has a thriving economy. Toward this end, payment rails are an important requirement to achieve a flourishing and fully realized Metaverse. At present, there is such enthusiasm for blockchains. What matters is that blockchains are programmable payment rails. That is why many position them as the first digitally native payment rails. Regardless of one's long-term belief in nonfungible tokens (NFTs), there are more interesting aspects of blockchain-based virtual worlds and communities. Tokens can be awarded for not only contributing time but also for good behavior such as community scores. Another question when designing a token is whether the token has an expiration date. Any fungible token might be programmed in a way that it expires after a certain date in order to prevent hoarding of the tokens. The idea of parallel money with an expiration date has a long history, e.g. Silvio

Gesell's "free money" (or *Freigold*) and regional currencies like the Austrian "Wörgl Schwundgeld" in the 1930s, which experimented with an inbuilt deflation of their currency, i.e. negative interest rate, to prevent not only hoarding but also inflation (see Chapter 6). It is interesting to note that Neal Stephenson's novel *Snow Crash* plays in the twenty-first century with hyperinflation created by the government due to the loss of tax revenue as people increasingly began to use electronic currency. Tokenomics will be instrumental in providing a system of incentives which enables players to earn tokens in the Metaverse. These play-to-earn and play-to-own activities to earn tokens in the Metaverse and create its own Metaverse economy. In addition to incentivizing transactions, the Metaverse should also provide gamified experiences. It is important to note that the Metaverse is more than Metaverse economy and blockchain games. There is a culture brewing here. Giving users something to do with items and building an activity around tokens offers experiences that give assets a purpose and create community, which is at the heart of the Metaverse economy. Thinking of the users solely as consumers will sell you short of the Metaverse's true potential. We have highlighted the unique potential benefits of the virtual world for society in that it provides a useful extension of the real-world economy by compensating for well-known market failures, e.g. rising income inequality. It is indeed a singular power of the virtual world that it can create anything for free, except labor. It is considered absolutely intolerable that a player have nothing to do. Thus, the virtual world must ensure that there is always another quest to do and that every player, at all times, has some way of turning her own action into some reward. As a result, unlike the real world, the virtual world has the potential to provide full employment by design. What is more, the virtual world can start all players at zero wealth and anyone who needs money can get it, since work is always available. Hence, all players start on an equal basis, giving rise to equality of opportunity. It has often been proposed that equality of opportunity ought to be the guiding principle of social policy. Policies that make the economic game obviously fairer are likely to become popular as virtual worlds broaden their influence. Moreover, what is striking is that even though online games exhibit economic inequality so vast that it dwarfs real-world inequality, nobody seems to care about inequality of outcomes. Indeed, it is more fun if the outcomes actually do differ wildly, because players expect that if one acquires some new power she should also be offered greater rewards. Otherwise, the virtual world would be no fun at all. In other words, people don't complain about a lack of vertical equity, but they howl about failed horizontal equity – not only in the virtual world but also in the real world.

■ **Overcoming Blind Spot #6: Important Shift from Passive to Active Experiences**

Ultimately, the Metaverse will be ushered in through experiences. Millions if not billions of users and dollars will be drawn to the new experiences and transformations that result. Shifting any of the time passively consumed thru today's lean-back entertainment to tomorrow's social, interactive, and more engaged entertainment is likely a positive outcome of the Metaverse. In Chapter 10, we have made the proposition that 6G, Next G, and the Metaverse should pave the way for the peak-experience machine that helps induce optimal states of consciousness by giving access to the upper range of human experiences, e.g. out-of-body experiences and afterlife experiences. The peak-experience machine aims at helping people not only have contact with a deeper, non-man-made metareality but also act upon Hannah Arendt's human condition characterized by fundamental activities, most notably work (ergon) and action in their surrounding environment of nature and others. After a new technology is created, society responds to it, which leads to new behaviors and new products, which in turn lead to new use cases for the underlying technology, thereby inspiring additional behaviors and creations. Large technological transitions often lead to societal change by tapping into widespread dissatisfaction with the present to pioneer a different future. Another important observation we have made is the fact that neither the Metaverse nor the Multiverse define any specific metrics (see Table 1.1). This is where the choice of using Society 5.0 as meta narrative will be instrumental in defining more complex metrics required for measuring the advancement of human capabilities as well as social mobility and equity.

11.2.3. Super Smart Society 5.0: From Galileo's Telescope to Metaverse's Telóscope

We have touched on the anticipated transition from today's technology-driven Industry 4.0 to tomorrow's human-centric Industry 5.0 and its two visions of human–robot co-working and a more holistic bioeconomy based on the two mutually beneficial principles of digitalization and biologization. These two principles are instrumental in creating a suitable framework so that economy, ecology, and society are perceived as necessary single entity and not as rivals. Almost every discipline shares promising interfaces with biology. In the long term, biologization will be just as significant as a cross-cutting approach as digitalization already is today. Industry 5.0 and Society 5.0 are two related concepts in the sense that they refer to a fundamental shift of our society and economy toward a new paradigm. Unlike Industry 5.0, however, Society 5.0 is not restricted to the manufacturing sector but addresses larger social challenges based on the integration of physical and virtual spaces. Society 5.0 counterbalances the commercial emphasis of Industry 4.0's focus on creating the smart factory. Conversely, Society 5.0 is geared toward creating the world's first super smart society. Whereas the main motivation behind Industry 4.0 and Industry 5.0 is mass production and a smart sustainable society, respectively, Society 5.0's main motivation is that humankind live in harmony with nature.

Embedding the Multiverse and its precursor Metaverse in the meta narrative Society 5.0 exploits the fact that the Multiverse, Metaverse, and Society 5.0 bear striking similarities (see Table 1.1). Given their underlying theme of fusion of digital and real worlds, it comes as no surprise that the Multiverse and Metaverse, embedded in the meta narrative Society 5.0, have many enabling technologies in common, ranging from virtual reality (VR)/augmented reality (AR), avatars, social robots to DAO, (nonfungible) tokens, and eudaimonic technology in support of delivering peak-experiences. Interestingly, we also observed an overlap of religious and spiritual dimensions, rooted in ancient knowledge, whereby Society 5.0 and the Metaverse are nothing new, but instead something quite ancient. Or, put differently, they are at once brand new and very ancient, ushering in the advent of a new culture in human history.

Advanced CPSS such as our proposed Internet of No Things, briefly mentioned in Chapter 1, will be instrumental in ushering in Society 5.0. It envisions Internet services to appear from the surrounding environment when needed and disappear when not needed. In Chapter 8, we have described the technical details of our CPSS based bottom-up multilayer token engineering framework for a *future stigmergic Society 5.0*, which enables humans to co-become supersmart by raising the collective human intelligence in a blockchain-enabled DAO, whereas at the same time letting them earn purpose-driven tokens as rewards. Toward this end, we borrowed ideas from the biological superorganism with brain-like cognitive abilities observed in social insect societies such as ants and bees. Specifically, we experimentally demonstrated the potential of the biological stigmergy mechanism for advancing collective intelligence in a CPSS-based DAO via tokenized digital twins. Note that biological stigmergy and blockchain technology have in common an inherent capability of self-organization and indirect communication by means of traces that members create in their environment. In both cases, society members record their activities in the environment using various forms of storage and use these records to organize and constrain collective behavior through feedback loops. Upon sensing these traces, other society members are stimulated to perform succeeding actions, thus reinforcing the traces in a self-sustaining autocatalytic way without requiring any central control entity. As a result, stigmergy maintains social cohesion by the coupling of environmental and social organization. Hence, with respect to the evolution of social life, the route from solitary to social life might not be as complex as one may think.

The future stigmergic Society 5.0 leverages on time-tested self-organization mechanisms borrowed from nature. In doing so, it follows the guiding principle of biologization. Biological mechanisms such as stigmergy hold promise to benefit from nature's efficiency, a process known as bioneering or biomimicry, which goes way beyond tackling energy efficiency in man-made systems. Given the urgent role of sustainability, future 6G and Next G networks should increasingly exploit nature-based solutions (NbS) found in nature-made systems, which involve working with nature to address societal challenges and provide human well-being in Society 5.0. The mutually beneficial symbiosis between biologization and digitalization is anticipated to create exciting opportunities and open up new research avenues in the coming 6G and Next G era. We must rapidly exit the current human-centered Anthropocene with its nonsustainability and enter the next era in human history named Symbiocene. We have seen that, beyond biomimicry, we must also have symbiomimicry such as the prime example of the so-called wood-wide-web, also known as the *Mycelium Internet*, which regulates forest ecosystems via fungal networks.

In Chapter 9, we have elaborated on how evolution makes us smarter and more social. We have seen that much of what makes humans biologically unique is their sociocultural experiences they create in socially shared realities. It is not sufficient to say that humans evolved to be smart or cooperative because being smart or cooperative is a generally good thing, so humans evolved to be smart or cooperative. Evolution does not work that way. Evolution is mostly conservative until a specific adaptive problem presents itself, e.g. exiting today's Anthropocene and entering tomorrow's Symbiocene. Then, those individuals best equipped to solve it have an adaptive advantage and so the species evolve. However, the fact that a psychological adaptation is aimed at a specific ecological challenge does note constrain its subsequent application. Remarkably, each of the past evolutionary transitions was characterized by a new form of cooperation with almost total interdependence among individuals and a concomitant new form of communication to support this cooperation, thereby creating a fundamentally new form of sociality. In all such cases, the basic structure was a we > me mode of operation in which "we" self-regulate each of us as individuals, as they gradually become fully fledged persons in a new culture.

The Metaverse should offer users novel tools to shape and behold human co-becoming in a future stigmergic Society 5.0 by leveraging on biological human uniqueness. We believe that a future stigmergic Society 5.0 provides a straightforward yet elegant solution for designing the future Metaverse by benefitting from evolution's capabilities to make us not only smarter but also more social, two attributes that are front and center in the super smart Society 5.0 vision. The symbiotic convergence of biomimicry (i.e. indirect communication via stigmergy) and advanced digitalization (i.e. purpose-driven tokens recorded in a blockchain-enabled online environment) represents a promising early example of symbiomimicry in the coming 6G and Next G era. For the Metaverse of the future, it is desirable to further cultivate the unique qualities that make us human, primarily social cognition via novel bifurcated experiences, which characterizes mature human thinking. Human reality is determined by humans' unique skill of social cognition. Unlike great apes and other animal species, humans are capable of bifurcated experiences. They understand that their perspectives – that is their beliefs, rather than truth – could potentially contrast with an objective (i.e. perspectiveless) view of the true nature of reality. This unique human capability of bifurcated experiences creates between humans a kind of shared world. This shared world could be the physical world, the realm of physical experiences through the age-old medium of life. Or more interestingly in the case of the future Metaverse, it could be a new kind of shared world arising from the fusion of digital and real worlds.

Section 9.3.2 gave us a good definition what reality and virtuality is in the first place. Ours is not a passive relationship, where reality is and we simply experience it. In fact, reality is a product of our minds, consisting of a constant stream of perceptions. Virtual reality is just an exercise in manipulating perceptions. In Section 9.4, we have dug deeper and asked whether cognitive science might also be helpful to better understand the true nature of reality in the age-old medium of real life rather than man-made virtual worlds. In particular, we explored answers to the questions whether we can trust our senses to tell us the truth and how they can nevertheless be useful even if they are not communicating the truth. We have seen that the external world actually consists entirely of a community or network of conscious agents that enjoy and act on experiences. The way one agent in a network perceives depends on the way that some other agents act, whereby conscious agents favor interactions that increase mutual information. Moreover, information, transacted in the currency of conscious experiences, is the fungible commodity of conscious agents, whose central goal is mutual comprehension.

According to cognitive science and its concept of conscious realism, evolution is hiding the true nature of reality in space–time from our eyes, i.e. space–time is like our own virtual reality. As a result, the things that we perceive don't exist independent of our minds. It is not simply that this or that perception is wrong. It is that none of our perceptions could possibly be right. Notwithstanding, there is an objective reality. Nevertheless, that reality is utterly unlike our perceptions based on our senses, transcending them. Metareality is not perceived by the five senses, yet it is totally accessible and offers our only means to escape simulated reality by becoming metahuman. Being metahuman is like tuning in to the whole radio band instead of one narrow channel. Note that many key ideas of modern cognitive science are in line with ancient Greek philosophers such as Plato and eastern religions such as Buddhism. In other words, they are at once brand new and very ancient, very similar to the original Metaverse and Society 5.0 vision.

Recall from Galileo's discoveries that it was not reason but a man-made instrument, the telescope, which actually changed the physical world view. Man had been deceived so long as he trusted that reality and truth would reveal themselves to his senses and to his reason if only he remained true to what he saw with the eyes of the body and mind. The human eye betrayed man to the extent that so many generations of men were deceived into believing that the sun turns around the Earth. Galileo was a polymath who pioneered the use of the telescope. It was the telescope, a work of man's hand, which finally forced nature, or rather the universe, to yield its secrets.

Now, wouldn't it be great if the future Metaverse would be a portal into metareality, which forces nature to yield its ultimate secrets of Being and transcend appearance beyond all sensual experience, even instrument-aided, in a re-found unity of the universe? Though man-made, the portal would help us peek into a new kind of shared world arising from the fusion of digital and real worlds beyond our five human senses, enabling us with extrasensory perceptions, sixth-sense experiences, and super-human capabilities to create the possibility of new kinds of concepts, including those that depend on an objective reality. Toward this end, the Metaverse should borrow from cognitive science not only the concept of token economy but also the aforementioned concept of conscious realism. In doing so, the Metaverse would give rise to an ingenious instrument that we may call the *telóscope*.

Recall from Chapter 1 that in order to ensure that Society 5.0 does not become a dystopian society, we have to redefine the modern concept of humanity and find a telós or purpose. The Metaverse, together with its underlying purpose-driven Web3 token economy and decentralized blockchain technology, should serve as the telóscope that enables humans to operate much like a hive mind, i.e. collective consciousness, where human labor will migrate into more deeply human spheres. Through the use of the telóscope, humans act as neurons in a human hive mind with blockchain technology acting as connective tissue to create virtual pheromone trails, i.e. programmable incentives, in novel dynamic media that exploit extended stigmergy mechanisms.

In doing so, we move headlong into the paradox of being simultaneously globalized and localized, into a more decentralized but interconnected world with a powerful connection to a greater intelligence coming from a source outside ourselves. By democratizing access to the upper range of human experiences via transformative technologies such as VR along with other technologies (see Section 10.4.4) that provide us a taste of transcendent communitas experiences, i.e. the fusion of the self and the world with increasing degrees of perceived unity, and contact with a deeper, non-man-made reality, modern-day Gutenbergs and Luthers are taking experiences once reserved for mystics and making them available for the masses, both for the perfection of the individuals themselves and, more importantly, for the perfection of the whole human community. Just as the original Renaissance retrieved the ancient Greek conception of the individual, our current renaissance retrieves the medieval and ancient understandings of the collective and reinvents the insights of premodern cultures in new forms, bringing us from individualism to something else that resonates with something greater than oneself.

Among possible true 6G innovations, Fettweis and Boche [2] claim that today's biggest challenge is the loss of trustworthiness and restoring trust must be understood as a basic societal challenge. They draw a historical comparison between today and the Renaissance, where Johannes Gutenberg's invention of the printing press in 1450 revolutionized society and heralded 300 years of renaissance. While Gutenberg's invention gave birth to printing, the Internet's full potential still remains to be unleashed in the years to come. It is well known that Gutenberg's printing press played a pivotal role in Luther's reformation of society. We don't yet know what the Internet truly is. Measured in Gutenberg time, we stand today at about the year 1481 with the progression of disruption in society. Note that Luther was born in the year 1483. Hence, the Internet's Martin Luther is yet to come.

11.3. Age of Discovery: Navigating the Risks and Rewards of Our New Renaissance

Recall from Section 10.4.5 that we might be in the midst of a renaissance. The apparent calamity and dismay around us may be less symptoms of a society on the verge of collapse than those of one about to give birth. We may be misinterpreting the natural process of birth as something lethal. Importantly, we saw that a renaissance without the retrieval of lost, essential values is just another revolution. A renaissance does not mean a return to the past. We don't go back to the Middle Ages. Rather, we bring forward themes and values of previous ages and reinvent them in new forms. Retrieval helps us experience the insight of premodern cultures that nothing is absolutely new. Everything is renewal.

This view is echoed by two award-winning Oxford scholars who redefine the present day as a new renaissance. In "Age of Discovery: Navigating the Risks and Rewards of Our New Renaissance," Ian Goldin and Chris Kutarna show how we can draw courage and wisdom from the last Renaissance in order to fashion our own golden age out of this new renaissance and reshape society [3]. They argue that a renaissance moment dares humanity to give its best just when the stakes are highest. Now, the same forces then converged 500 years ago to spark genius and upend social order – great leaps in science, technology, education, and trade, among others – are once again present, only stronger and more widespread.

According to [3], in the Renaissance, the polarizing effects of progress were clear to see. While average welfare improved through much of the period, the margins of society, rich and poor, grew further and further apart. As trade and new forms of manufacturing expanded, the gulf between rich and poor widened. By 1550, in nearly every sizable Western European town, the top 5–10% of residents owned 40–50% of total wealth, while the bottom 50% owned little more than their own labor. A major cause was falling real wages at the bottom of the income scale, especially among the unskilled. By 1550, youth unemployment was common. The widening gap between rich and poor betrayed the limits of the period's high-minded idealism. Humanists exalted "Man" but many seemed to ignore the squalid condition of ordinary men. Fast forward to present day, while average global welfare has improved, the extremes have spread further apart, so that today the top and bottom live in ever-sharper contrast to each other. In 2010, the 388 richest billionaires in the world controlled more wealth than the bottom half of all humanity. In 2015, it took just 62 people to make the same claim. The bottom half of the world's population – 3.6 billion people – live, on average, on just a few dollars per day. Importantly, while the difference *between* developing and developed countries has converged over the past quarter-century, the difference *within* countries – both developing and developed – has diverged. Within almost all countries, from the least developed to the most, the gap between rich and poor has widened over the past few decades. Even countries long known for income equality, such as Denmark, Germany, and Sweden, have seen the rich pulling further away from the pack.

11.3.1. The Formula for Mass Flourishing

Public frustration with economic injustices spoiled much of the last Renaissance and could do the same in the new renaissance. Goldin and Kutarna [3] argue that Europe's leap forward during the Renaissance suggests that the presence of great, focused minds was not itself sufficient for genius to erupt society-wide. Something else mattered: *collective genius*. Every person possesses a unique fragment of capability. Collective genius happens when society nurtures and connects those diverse fragments. The ability to creatively combine past solutions with present technical problems was highly prized and spreading fast. Mainz, Gutenberg's hometown, was a crossroads for two very different domains: *wine-making* and *coin-minting*. These critical crafts were deep and local, but once Gutenberg successfully combined them, the diffusion of his press was guaranteed by the more general forces of the era (despite his own efforts to keep the technology secret).

Goldin and Kutarna [3] further elaborate on the question of what *environmental conditions* make collective genius flourish in some times and places, and not others. The following three conditions in particular made fifteenth- and sixteenth-century Europe ripe for a *collective heyday*, and which today's scholars consider still decisive:

1. The first condition was a jump in the velocity, variety, and richness of the flow of ideas. It is an obvious, but essential point: the more quickly ideas flow, the more rapidly new and fruitful combinations of ideas can emerge. Variety matters too because, as Gutenberg found and contemporary research confirms, the big leaps tend to happen when seemingly unrelated domains collide. Moreover, the richer the flow, the more complexity it can carry. The most direct catalyst of this enhanced flow of ideas was the new medium of print. It multiplied the available body of knowledge and broadened the network of practitioners in every field. Engaging greater numbers of people brought a wider set of knowledge, experiences, and ideas to every important problem. Then and now, new exchanges between civilizations and social mobility are of critical importance. One of the best ways to push the limits of our present thinking is to meet people who think differently.

2. The second condition that helped collective genius flourish was a booming stock of well-educated, well-fed brains to tap this idea flow: failed attempts that discover dead ends so that others don't have to; technical tweaks to devices and instruments that make it possible to probe deeper into a mystery; countless debates, oral and written, with other people's viewpoints to firm up one's own grasp of a craft or surprise the mind into new directions. The more brains that understand a craft and grapple with its limits, the more likely that someone will surpass those limits.

3. The third condition that helped collective genius flourish more strongly in Renaissance Europe than in other places was strong private and social incentives to reward risk-taking. Europe was comprised of many, small states. Competition and war between them urged each to invest in discoveries that might offer an edge. Wealthy cities

endowed new schools, universities, and professorships to tackle important commercial problems. Meanwhile, rich families pumped money into new art, sculpture, and architecture. It helped the emerging class of wealthy merchants gain a veneer of gravitas, and it was one of the few socially acceptable forms of ostentation. On the supply side of this emerging marketplace for ideas, individuals who offered good ones were free to profit from them. Gutenberg's wunder-machine was copied so quickly and widely that he made no money from it, but the generation of inventors after him fared much better.

According to [3], the same conditions are present now, only stronger and more widely felt. New general-purpose technologies (GPTs) have once again made mass communication cheap and abundant, creating new idea-rich contact points that have multiplied between different peoples, products, and lifestyles. Migrants spread and connect the world's brainpower. The world suddenly contains more brains, exchanging an exploding volume and variety of ever-more vivid ideas, globally, instantly, and at near-zero cost. These conditions describe an ideal world for creative breakthroughs, both individual and collective. They are why big shifts are happening now, and why the genius of this new renaissance ought to far surpass the last. If we get the incentives right (e.g. purpose-driven tokens in a Web3 token economy based on decentralized blockchain technologies), our present flourishing will only get bigger.

11.3.2. Cognitive Blind Spots: Social Complexity vs. Cognitive Capacity

Goldin and Kutarna [3] make the point that complexity can bring several benefits. It increases the number and variety of good things that can touch us and that we can reach out to. Further, it is a major catalyst of creativity and idea generation. From a risk perspective, too, complexity can be a good thing. The greater variety and volume of connections and flows create redundancy, of which the Internet is contemporary life's best example. When one link goes down, its traffic re-routes almost instantaneously to alternatives, so that our end-user experience is not often interrupted. Clearly, complexity breeds benefits.

However, they argue that complexity also presents a problem. The more complex our interactions become, the harder it is for us to see relationships of cause and effect. We develop cognitive "blind spots" in our vision of the events around us when social complexity rises faster than our cognitive capacity for understanding, as illustrated in Figure 11.1. The absence of any clear understanding of cause and effect leaves a cognitive gap that society stuffs full of stereotypes, superstitions, and ideological agendas. Rising complexity in our

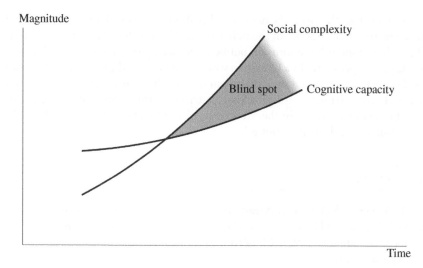

Figure 11.1 We develop cognitive "blind spots" when social complexity rises faster than our cognitive capacity. Source: Goldin and Kutarna [3].

social systems strains our cognition. It is the other side of the tangled, rapidly developing age we have been born into, and it permeates our lives, including our relationships with one another. With complexity, the hardest part of solving a problem is seeing it.

Goldin and Kutarna [3] conclude by mentioning Edmund Phelps's concept of *mass flourishing*, which we have introduced in Section 9.2, where we postulated that the Metaverse should become a platform that enables human mass flourishing. In line with Phelps, Goldin and Kutarna [3] urge their readers to replace a palpable decline in vitalism – in doing – with more experimentation, exploration, tinkering, and guessing, so that we as individuals and society will not merely plod along, but thrive. In an age of discovery, the balance between risk and reward tips in favor of taking bold action. First, because there is suddenly so much new territory to explore. Second, because our entanglement accelerates and spreads the value we can create for ourselves and others. Third, the costs of failure are plummeting due to increasing digitalization. The first courageous act is to take a long-term, big-picture view. Set out on a new voyage of discovery. Columbus sought Asia and found America. Likewise, this century will reward explorers who set out on paths whose final destination is uncertain. This has always been true, but the opportunities are rarely as rich as they are right now, gaining unique, defining experiences and incentivizing us to seize the opportunities in front of us. The lure of deep craft was very strong then, but it is much stronger now. The range of disciplines that must come together to make not only a breakthrough is wider but also demand more from the rich. Some of the rich do recognize the debt they owe to society. The Medici spent huge sums on public works partly out of generosity but also under social pressure. In a Christian world that professed the virtues of charity and poverty, possessing huge wealth put one on shaky moral ground. The rich needed a new virtue to legitimize their outsized fortunes, and patronage was it.

Find your Florence in the Metaverse, physically, virtually, or best both. In the Multiverse, think *and*, not *or*!

References

1. T. Winters. *The Metaverse: Prepare Now For the Next Big Thing!* November 2021.
2. G. P. Fettweis and H. Boche. 6G: The personal tactile internet - and open questions for information theory. *IEEE BITS the Information Theory Magazine*, 1(1):71–82, September 2021.
3. I. Goldin and C. Kutarna. *Age of Discovery: Navigating the Risks and Rewards of Our New Renaissance*. St. Martin's Press, May 2016.

EPILOGUE

Portrayal of the Renaissance:
Kairos and Metanoia.

T he conventional portrayal used through the original Renaissance shows
Kairos, the god of opportunity, and goddess Metanoia in his wake. The
ancient Greeks had two words for time: Chronos and Kairos. The former stands for our typical quantitative usage of the word, i.e. chronological

6G and Onward to Next G: The Road to the Multiverse, First Edition. Martin Maier.
© 2023 The Institute of Electrical and Electronics Engineers, Inc.
Published 2023 by John Wiley & Sons, Inc.

time that marches forward. The latter is more qualitative in nature, representing an opportune moment in time, frequently characterized as Kairos, the god of opportunity, in Greek mythology. The goddess Metanoia accompanies him, sowing regret and sorrow that the opportune moment had not been grasped. But she also holds out her own opportunity: a fundamental transformation of man's vision of the world and of himself, a transmutation of consciousness, a *change of mind*, which is how metanoia is literally translated from Greek *meta* (beyond) and *nous* (mind). So, what regrets might we have in the future if we do not change the way we think and act in our new renaissance in the coming 6G and Next G era?

Tackling the Impossible: Extreme Innovation in Matter and Mind. Where the original Renaissance brought us the printing press, our era brings the Internet and a newfound respect for networked intelligence and connectivity to achieve a networked collective consciousness. The Internet is still in its infancy, it's still a work in progress. So are our business models and ethical frameworks. We are still vested with the human agency and responsibility to design and steer the net we wish to see in the future. In his latest book *The Art of Impossible: A Peak Performance Primer*, Steven Kotler makes the case that we are capable of so much more than we know. He defines impossible as a kind of extreme innovation that exceeds both our capabilities and our imagination, whereby those who tackle the impossible are not just innovating in matter but also in mind. As humans, we have all been shaped by eons of evolution. Biology is the very thing designed by evolution to work for everyone, i.e. biology scales. Kotler argues that basic neurobiological mechanisms shaped by evolution, present in most mammals and all humans, are beneath the art of impossible and whose biggest impacts are cognitive. In fact, he explains that scientists had made some serious progress in this arena. Experiences that were once seen as "mystical" were starting to become known as "biological." Interestingly, Kotler points out that many ancient cultures, including the Greeks, Indians, and Chinese, thought of creativity as discovery, because ideas came from the gods and were merely discovered by mortals. This shifted during the Renaissance, when insights bestowed by the divine became ideas kindled in the minds of great people.

And this brings us to our final question: How can we regain access to this higher realm of mystical or biological experiences? Kotler provides some useful hints how to enter the unitary continuum with increasing degrees of perceived unity between self and world. Among other flow triggers, a rich environment is crucial. This is a combination platter of three separate triggers: (i) *novelty*, (ii) *unpredictability*, and (iii) *complexity*. Novelty is one of our brain's favorite experiences. Novelty could mean that there's either danger or opportunity lurking in our environment. Unpredictability means that we don't know what happens next. Thus, we pay extra attention to the next. For illustration, Kotler mentions a study at Stanford University that shows that the dopamine spike produced by unpredictability, especially when coupled with novelty, comes very close in size to the spike produced by substances such as cocaine. It's a nearly 700% boost in dopamine, which leads to a huge boost in focus. Complexity shows up when we force the brain to expand its perceptual capacity, e.g. when we gaze up at the night sky and realize that a great many of those singular points of light are actually galaxies. This is the experience of awe, where we get so sucked in by the beauty and magnitude of what we're contemplating that time slows down and the moment stretches on into infinity.

In addition, Kotler highlights the importance of *social triggers* that trigger the shared, collective experience of hive mind where people drop into flow together. For group flow to arise, everybody needs to be heading in the same direction. Shared, clear goals is how this happens. This doesn't have to be fancy. What matters is that the group feels like they are moving together toward the same (or complementary) targets. One of the most well-established facts about flow is that the state is ubiquitous. It shows up anywhere, in anyone, provided certain initial conditions are met. There has been considerable work done on both the nature of this shared flow experience and, at least from a psychological perspective, what might be causing it. Yet, technological limitations have stood in the way of deeper research of group flow or group flow's triggers. Or as Kotler puts it, there are still gaps in the science you could drive a bus through.

We have seen in this book that the Metaverse should become a platform that enables human mass flourishing, whose wellspring is modern values such as the desire to create, explore, and meet challenges. Humans' desire to create, explore, and meet challenges can be best realized in a future Metaverse where there is so much new territory to explore. The Metaverse is uniquely positioned to be loaded with the aforementioned important triggers of novelty, unpredictability, and complexity as well as social triggers via shared, collective experiences. The Metaverse offers users novel tools to shape and behold human co-becoming in a future stigmergic society by leveraging on biological human uniqueness. In particular, the Metaverse is supposed to cultivate the we > me mode of operation of humans by developing the unique qualities that make us humans and create between them a kind of shared world. This shared world could be a new kind of shared world arising from the fusion of digital and real worlds to create the possibility of new kinds of concepts, including getting access to metareality. The future Metaverse should be a portal into metareality. Though man-made, the portal would help us peek into a new kind of shared world arising from the digital and real worlds, beyond our five human senses. Modern research suggests that the human capacity of self-loss occurring during such transcendent experiences is the portal to many of life's most cherished experiences that often feel realer than real.

The Road to the Multiverse: The Road to Eleusis 2.0. In ancient Greece, Athens' best and brightest flocked to Eleusis for 2000 years. Eleusis was located on the outskirts of ancient Athens. Mysteries, from the Greek verb *muo* (*μυω*) meaning "to shut one's eyes," were said to hold the entire human race together. Plato was permanently transformed by whatever he observed in Eleusis. Aristotle is known for saying that "initiates came to Eleusis not to learn something, but to experience something." Wouldn't it be nice if the Multiverse became the new *Eleusis 2.0* – a parallel plane on the brink of the Internet for human leisure, labor, and existence more broadly – where we "shut" our eyes using advanced extended reality (XR) head-mounted devices, which give way to awe-inspiring transcendent experiences that are at once brand new and very ancient?

Giordano Bruno, who was burned at the stake in the Campo dei Fiori in Rome in the year 1600, conceived of a Multiverse centuries before the idea was seriously regarded by theoretical physicists of the twentieth and twenty-first centuries. Bruno's thought stood between the old and new eras. It was both ancient and modern, similar to the Metaverse, the next step after the mobile Internet, which stands between the past generations of mobile networks and the coming 6G and Next G era and future Multiverse.

Giordano Bruno's Multiverse:
A glimpse of his many worlds.

INDEX

6G and Onward to Next G: The Road to the Multiverse, First Edition. Martin Maier.
© 2023 The Institute of Electrical and Electronics Engineers, Inc.
Published 2023 by John Wiley & Sons, Inc.

THE COMSOC GUIDES TO
COMMUNICATIONS TECHNOLOGIES

Nim K. Cheung, *Senior Editor*
Richard Lau, *Associate Editor*

The ComSoc Guide to Next Generation Optical Transport: SDH/SONET/OTN
Huub van Helvoort

The ComSoc Guide to Managing Telecommunications Projects
Celia Desmond

WiMAX Technology and Network Evolution
Kamran Etemad and Ming-Yee Lai

An Introduction to Network Modeling and Simulation for the Practicing Engineer
Jack Burbank, William Kasch, and Jon Ward

The ComSoc Guide to Passive Optical Networks: Enhancing the Last Mile Access
Stephen Weinstein, Yuanqiu Luo, and Ting Wang

Digital Terrestrial Television Broadcasting: Technology and System
Jian Song, Zhixing Yang, and Jun Wang

TV White Space: The First Step Towards Better Utilization of Frequency Spectrum
Ser Wah Oh, Yugang Ma, Edward Peh, and Ming-Hung Tao

Digital Services in the 21st Century: A Strategic and Business Perspective
Antonio Sanchez and Belen Carro

Toward 6G: A New Era of Convergence
Amin Ebrahimzadeh and Martin Maier

Printed and bound by CPI Group (UK) Ltd, Croydon, CR0 4YY

27/10/2024

14580672-0001